Statistical Forecasting

Statistical Forecasting

Warren Gilchrist

*Head of Department of Mathematics and Statistics,
Sheffield City Polytechnic*

A Wiley–Interscience Publication

JOHN WILEY & SONS

Chichester · New York · Brisbane · Toronto

Library of Congress Cataloging in Publication Data:

Gilchrist, Warren Garland, 1932—
Statistical forecasting.

'A Wiley—Interscience publication.'
1. Time-series analysis. 2. Prediction theory.
I. Title.
QA280.G54   519.5'4   76-13504

ISBN 0 471 99402 2 (Cloth)
ISBN 0 471 99403 0 (Paper)

Typeset in IBM Century by
Preface Ltd, Salisbury, Wilts
and printed in Great Britain by
The Pitman Press, Bath

Preface

From the dawn of recorded history, and probably before, man has sought to forecast the future. Indeed, the ability to foresee the consequences of actions and events is one of the defining properties of 'mind'. Where the phenomenon being forecast was a purely physical one, such as the occurrence of midsummer's day or of an eclipse, man was able from very early times to obtain very accurate forecasts. Initially such forecasts were derived on a purely empirical basis. Methods were found that worked without any basic understanding as to why they worked. Later, as greater understanding of the phenomena developed theoretical models were developed which enabled forecasts to be obtained on a more reasoned basis. During the last century interest focused on a number of phenomena, such as economic variation and sunspot cycles, where

(a) there were available series of observations taken over a period of time, called time-series, upon which forecasts could be based;
(b) purely mathematical rules were found to be inadequate to describe the phenomena since they involved features of a chance nature.

Over the last fifty years approaches to this type of problem have been developed which seek to allow for the influence of chance and for the unknown complexity of these situations. Thus, forecasting methods were developed which were essentially statistical in nature. These were based on using statistical techniques to fit models of a wide variety of types to past time-series data and on forecasting the future from these models.

The aim of this book is to provide the reader with an understanding of the methods and practice of statistical forecasting. The structure of the book is based on the structure of the activities involved in the practice of statistical forecasting. Figure P.1 summarizes this structure. Part I of the book provides a general introduction to some of the basic concepts and terms in statistical forecasting. Part II deals with the development of statistical forecasting methods applicable to each of a variety of different situations. The approach to classifying this part into chapters has been to classify the situations; e.g. one chapter considers

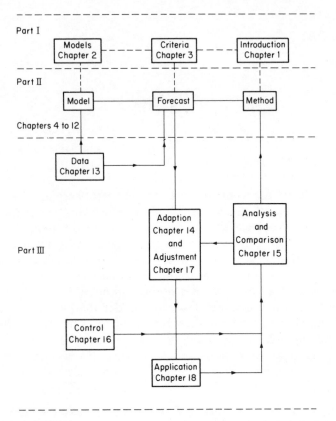

Figure P.1   The structure of the book and of statistical forecasting

trending time series, another seasonally varying time series. We then consider the various methods that might be used in each situation. In practice there is considerably more to statistical forecasting than putting data into a formula to obtain a forecast. Part III deals with these further facets. In Chapter 13 we look at the data that is needed for forecasting, where it might come from and how it may be treated. Chapter 14 looks at ways in which the forecasting methods of Part II may be adapted to apply to a wider range of applications and to give a better quality of forecast. In Chapter 15 we examine a wide range of methods for analysing and comparing forecasting methods. Chapter 16 has the same basic aim, but is conerned more with the routine control of an operational forecasting system. In Chapter 17 we use some of the properties of forecasts, that might be identified using the methods of Chapters 15 and 16, to create even better forecasts. Always we seek to bear in mind that forecasts are produced for practical purposes that may influence the way we obtain and use our forecast in the first place.

This interrelation between forecast and application is explored in Chapter 18.

The general aim of the book is to give a systematic account of the methods of statistical forecasting in common use and of the practical aspects of their use; consequently, the level of mathematics and statistics assumed of the reader has been kept to that which the writer's teaching experience suggests as appropriate to people whose interest is in obtaining forecasts for practical purposes. This is mainly limited to the basic ideas of statistics and the ability to follow simple algebraic arguments. Where more than this is used, the section is marked with an * and can be omitted without detriment to the understanding of later sections. Where certain ideas are used that might be unfamiliar to a significant proportion of readers, these are discussed in the appendices. Wherever possible, the discussion of each forecasting method is developed and illustrated by example in such a way that the reader may then be able to use the method himself. There are, however, a number of methods whose mathematical development has been regarded as beyond the scope of this book, but computer programs are commonly available for their use. The author's aim in these cases has therefore been to discuss and illustrate the basic concepts of the methods. Thus the reader should be able to use such computer programs with understanding of their principles though not of their technical details.

A short list of relevant references which are referred to in the text are given at the end of each chapter.

Acknowledgements

I would like to express my thanks to the many people who have discussed forecasting problems with me over the years and those who have given me comments on early drafts of this book. In particular, I would like to thank N. Booth, O. Oliver, V. G. Gilchrist and also numerous students who have helped me develop suitable approaches to many of the topics in this book. Finally, I would like to thank the Social Science Research Council for their support of work on the analysis of forecasting methods that underlies, particularly, Chapter 15.

W. G. GILCHRIST

Contents

Notation

x_t an observation at time t

$\hat{x}_t$ a forecast, or estimate, of x_t

$\tilde{x}_{t,h}$ a forecast made at time t of x_{t+h}, i.e. with lead time h

$\tilde{x}_t$ the one-step-ahead forecast abbreviated from $\tilde{x}_{t,1}$

e_t the forecast error $x_t - \hat{x}_t$, often $x_t - \tilde{x}_{t-1}$

$e_{t,h}$ the lead time h forecast error $x_{t+h} - \hat{x}_{t+h}$

μ_t the underlying mean at time t, i.e. $E(x_t)$

β_t the underlying trend at time t

Part I

Preliminaries

Chapter 1

An introduction to forecasting

1.1 Everybody forecasts

Every time we make an appointment we are making a forecast about our future ability to keep that appointment. At the 'see you in ten minutes' level we do not even think about the forecast, though for more long-term appointments we try to think of any obvious things that might stop the meeting from occurring as forecast. The more we think about our forecast, the more likely it is to be right, for we will not be so likely to overlook some prior engagement or some other aspect of the situation. In this example we might even try to assess our confidence in our forecast by the amount of surprise we would show when the arrangement falls through. Such acts of unconscious or semiconscious forecasting are part of our everyday lives; they are also part of the lives of those who are involved with industrial or business problems. The good manager is not so much one who can minimize the effects of past mistakes but rather one who can successfully manage the future. Consider, for example, the following list of questions:

What will next month's sales be?
How much should be produced this month?
How big a stock should be kept?
How much material should be bought?
When should the material be bought?
What should the sales target be?
Should the labour force be increased?
What price should be charged?
How many operators are required?
What will the profit be?

To obtain the best anwers to these questions, we would need to be able to see into the future. To be able to give the best practical answers to these questions, we need to be able to forecast the future. Our forecasts may be wrong, but they must be made. In the past most answers to questions of the above type have been based on unconscious or semiconscious forecasts. In these forecasts it was frequently assumed that the future would be just like the recent past. The growing interest in many aspects of forecasting is based on the belief that conscious and careful thought cannot fail to help improve our forecasting skill and thus our ability to get answers to the above types of question. It may be in some cases that a thorough scientific forecasting study might not lead to much better forecasts than the old 'by guess and by gosh' methods. Even so, such a study will enable those involved to have a better understanding of their situation and thus improve their control of it

There are many methods of statistical forecasting, none of which can be put forward as providing all the answers to everybody's forecasting problems. The techniques discussed in this book are chosen because they have found the widest application. The emphasis of the book, however, is not on specific techniques but on general methods and attitudes towards forecasting. In this way it is hoped that the reader will be better able to apply statistical forecasting methods to his unique problems. New techniques and applications of forecasting are being published every month, and thus if this book is regarded simply as a compendium of forecasting techniques it would very soon become obsolete. If, however, the reader treats it as a text on general methods and attitudes in statistical forecasting, then he should be able to incorporate the new developments in this rapidly expanding field into his basic framework of knowledge.

1.2 Methods of forecasting

In this section we shall define three general methods of forecasting that involve very different approaches. It is not intended to imply that these are the only types available, or that any specific method must automatically be classifiable as one of these, but these are introduced to provide a general background.

1.2.1 Intuitive methods

These are the classical methods of forecasting. They are based essentially on the individual's feeling for the situation, or on the expert's opinion. Often one is forced into these methods by the complexity of the situation and the sparceness of reliable and relevant data. One obvious way of seeking to improve these methods is to replace the single expert by a small group or larger number of people. For example, in forecasting sales one might use the summarized views of the sales force or of a committee of the area sales managers.

Alternatively, a sample survey of the views of potential customers might yield more reliable information.

1.2.2 Causal methods

These methods are based on trying to forecast effects on the basis of knowledge of their causes. In many forecasting situations the causes are economic and hence most of the work in this field is also of an economic nature. Economic relationships between causes and effects are studied; as the effects occur after the causes direct knowledge of the causes leads to a forecast of the effects. But if, as often occurs, the time lag between cause and effect is very small, we may still have to forecast the causes, since information on the causes may not become available until some time after the effects have occurred. The causes used for forecasting are then chosen as those that can be forecast in a more reliable fashion than the effects. If the mechanism of the relationship is known, quite elaborate forecasting methods may be devised. Often, however, the nature of the relationship is obscure and we are forced to use some purely empirical form of relationship. Chapter 6 gives a discussion of some of the relevant forecasting methods. There are a considerable number of books dealing with economic forecasting; the references list some examples.

1.2.3 Extrapolative methods

These methods are all based on the extrapolation into the future of features shown by relevant data in the past. The methods are usually of a mathematial or statistical nature and provide the basis for most of this book. Such methods are sometimes referred to in books on economic forecasting as naive methods. However, to understand any situation we must start by being naive, and consider the simplest methods and approaches. When we understand these thoroughly, we may then progress to the higher realms.

1.3 The basis of forecasting

To attempt a scientific study of forecasting the future might seem a hopeless exercise, since each problem is obviously unique. As we look more closely at specific examples of forecasting, however, we find some common priniciples. Let us consider some examples.

Example 1.1

Most political policy-makers are concerned with forecasting the future as it would be if no new actions were taken. They then consider

6

the *structure* of society as they see it and take action to introduce policies that will improve society according to their political criteria. To be able to decide what action to take, they must be able to identify structures in society that will respond to their policies. It is not unknown for the implemented policies to fail. This might be due to an incorrect identification of the structure or, alternatively, to a lack of *stability* in that structure. To correctly make such forecasts and define policy, the policy-maker needs to be able to identify the appropriate *structure* and have some confidence in its *stability* over the appropriate period.

Example 1.2

H. T. Davis (1941) points out that had the mass of the sun not been so great, relative to the planets, statistical methods would have had to be used to investigate the physical laws of gravitation and to predict future movements of the planets. As things are, the gravitational pull of the sun on a planet is so powerful that the planet's orbit is almost the same as it would be if the other planets did not exist. From observations on this orbit the *structure* of the relationship between sun and planet can be inferred and the inverse square law of gravitation deduced as the model that fits most of what is observed. The great *stability* of the situation enables very accurate forecasts of the future positions of the planets to be made. Had the mass of the sun been smaller, its effects would not have dominated those of the other planets and much more complicated orbits would have occurred. The consequence of this complication would have been that, though the inverse square law would still be the basic law, it would have been much harder to find it from the observed data. Further, forecasting the future positions of the planets would have been much more difficult, as the simplicity and stability of orbit we have would not have occurred.

Example 1.3

As a third example we will consider some of the forms taken by a set of data which might correspond to orders received by a firm each week. Table 1.1 gives the appropriate data. Suppose the form of the data is that given in line (a) of the table. Here the orders come from just one customer whose demand is constant. The data show a very simple *structure*, constant demand, and a perfectly *stable* behaviour. We would thus feel fairly confident in forecasting orders of 6,000 items for week 13. Suppose, however, the data took the form shown in line (b) of Table 1.1. Here the occurrence of 9,000 units of demand at week 5 leads to some doubt about the stability of the structure shown in the remaining weeks. Hence we might still forecast 6,000 items for the next

Table 1.1 Orders received (in 000s)

| | | | | | Week number | | | | | | |
	1	2	3	4	5	6	7	8	9	10	11	12
(a)	6	6	6	6	6	6	6	6	6	6	6	6
(b)	6	6	6	6	9	6	6					
(c)	6	6	6	6	9	6	6	6	6	9	6	6
(d)	7	6	6	4	7	6	8	6	5	6	7	5
(e)	8	8	9	8	12	12	15	14	14	16	18	17

week, week 8, but no longer with such confidence. Suppose that we wait a few more weeks to see what happens and enquire into the reason for the demand for 9,000 units. We may find that the data in (b) arise from orders by two customers, the one referred to above and also another customer who orders 3,000 items at regular five-weekly periods. The sensible forecast now would be 6,000 items for weeks 13 and 14 and 9,000 items for week 15. We see then that it is sometimes possible to cope with what appears to be an instability in a structure by considering more sophisticated forms of structure. The apparent instability of data (b) becomes the stable structure when more information is obtained, as in data (c).

In practice, data such as those in (a), (b) and (c) do not occur very frequently. We are much more likely to see data such as those shown in line (d) of Table 1.1. These data do not show the nice simple mathematical structure of (a), but nonetheless they do show a structure, as is seen from Figure 1.1. This gives a simple bar chart of the data, showing the frequency with which the sizes of orders occur, and it reveals a quite clear structure, but in this case it is a statistical structure rather than a mathematical one. It is apparent that the average order is near to 6,000 units, and, on the evidence of this small sample of data, the orders appear to be distributed fairly evenly about this value with decreasing frequencies of occurrence away from the average.

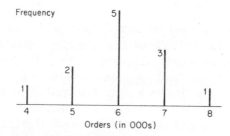

Figure 1.1 Bar chart of the data of Table 1.1(d)

8

The structure of this situation is thus a structure in the probabilities of orders taking on certain values. To describe such a structure we will need the language of statistics as well as mathematics. When the term 'stability' is applied to this situation, it must be applied to the whole statistical structure. The values of the orders are not stable in the mathematical sense, as they vary from week to week in a chance fashion. What is assumed to be stable is the structure of this chance behaviour as illustrated for the sample of 12 weeks by Figure 1.1. In forecasting the orders for week 13 it would be reasonable to forecast 6,000 units, but now we are not completely certain of our forecast as we were for data (a) and we are not subject to vague doubts as we were for data (b). The information in Figure 1.1, i.e. the statistical structure, enables a much clearer idea to be formed of the possible errors in the forecast. If we take the data as providing estimates of the probabilities of error, and consider only units of thousands, we have a probability of about $\frac{5}{12}$ of being right with our forecast of 6,000, about $\frac{5}{12}$ of being 1,000 units in error and $\frac{1}{6}$ of being 2,000 units in error.

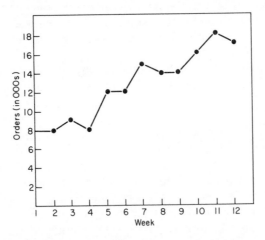

Figure 1.2   Plot of the data of Table 1.1(e)

Consider, finally, the set of data in line (e) of Table 1.1 which is plotted in Figure 1.2. It is clear from the graph that the data lie in a 'chance' fashion about a line. This might occur if, added to the source of the orders in data (d), and additional customer ordered 1,000 units in week 1, 2,000 units in week 2, and so on. Thus in mathematical terms his orders in week n are 1,000 n. It is thus clear that the structure of the data has both a mathematical and a statistical aspect. To say that such a structure is stable, we imply that both the upward trend and the statistical variation about that trend are both stable. The forecast for week 13 would be our forecast of 6,000 units as made for week 13 in section (d) plus the 13,000 units from the trend, the new customer, giving 19,000 units. The knowledge of the statistical structure would enable exactly the same probability statements to be made as were made for the forecast for data (d).

Example 1.4

Table 1.2 gives two years' data and Figure 1.3 gives a plot of this set of data. At first sight these data look unstructured. A study of the situation in which it was obtained suggests that it might be composed of an upward trend, some sort of seasonal variation and a chance variation, or as it is usually called, a random variation. Table 1.3 shows a breakdown of the data into these three elements, together with a table of the frequencies of the random variation, which gives the same sort of information as Figure 1.1. Thus the data have a structure composed of both mathematical and statistical elements. It is clearly seen from

Table 1.2 Two years' data

Year	Month											
	1	2	3	4	5	6	7	8	9	10	11	12
1	33	34	30	36	26	22	14	25	19	20	38	37
2	42	49	42	33	37	34	23	40	23	31	44	40

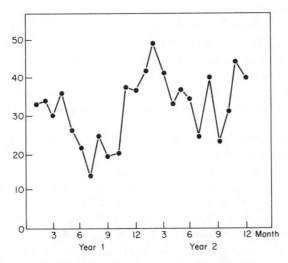

Figure 1.3   Plot of the data of Table 1.2

Table 1.3 Breakdown of data in Table 1.2

(a)

Month	Trend Year 1	Year 2	Seasonal additions	Random variation Year 1	Year 2	Total Year 1	Year 2
1	21	33	10	2	−1	33	42
2	22	34	12	0	3	34	49
3	23	35	10	−3	−3	30	42
4	24	36	8	−1	−11	31	33
5	25	37	0	1	0	26	37
6	26	38	−6	2	2	22	34
7	27	39	−16	3	0	14	23
8	28	40	−10	7	10	25	40
9	29	41	−9	−1	−9	19	23
10	30	42	−7	−3	−4	20	31
11	31	43	2	5	−1	38	44
12	32	44	6	−1	−10	37	40

(b) Frequency table for random variation

Values	Less than −7,	−7 to −5,	−4 to −2,	−1 to 1,	2 to 4,	5 to 7,	8 to 10
Frequency	3	0	4	9	5	2	1

this example that finding the underlying structure for a set of data can be a difficult task. It is certainly not obvious that the data of Table 1.2 were obtained in the fashion shown.

It is also clear that, given enough effort and ingenuity, one could create some such structure for any set of data. If, however,we are going to assume stability of the structure for forecasting purposes, it is clear that the structure that is used must correspond to some real structure in the situation. The mathematical and statistical equations used in forecasting must make sense in terms of the situation in which they are used. The search for a structure for use in forecasting is thus not simply a mathematical or statistical task but requires a broad knowledge of the nature and background of the particular forecasting problem being investigated.

We have seen from the above examples that the basic requirement for forecasting is the existence of a stable structure. This structure may be of purely mathematical form or purely statistical form or, most frequently, a mixture of both. So far these structures have been dealt with in a purely descriptive fashion; to proceed further we must be more precise and rigorous in the description of the structures. Mathematical and statistical models of forecasting situations provide such a rigorous description. It is an introduction to the study of such models that provides the subject matter for the next chapter.

References

Chisholme, R. K., and Whitaker, G. R. (1971). *Forecasting Methods.* R. D. Irwin, Homewood, Illinois.

Davis, H. T. (1941). *Analysis of Economic Time Series.* Principia Press.

Morrell, J. (1972). *Management Decisions and the Role of Forecasting.* Penguin, Harmondsworth.

Robinson, C. (1971). *Business Forecasting. An Economic Approach.* Thomas Nelson and Sons, London.

Chapter 2

Models

2.1 Scientific forecasting

It has been seen in Chapter 1 that forecasting is an inherent part of business and industrial activity and that until recently it has been done mainly by intuitive methods. An essential part of the present revolution in business methods is the attempt to use scientific methods as an aid to making business decisions. Let us consider now some of the steps which are involved in applying scientific method to forecasting problems.

2.1.1 The first step — data collection

Before we can apply any scientific forecasting method to a situation, we must naturally examine the situation in great detail. In particular, this involves gathering together as much relevant data on the situation as possible. Such data are of two main types: firstly, data which are internal to the firm, e.g. records of past sales, production, stocks, etc.; secondly, data which are external to the firm. For example, Government and Trade Association statistics often provide a background against which to interpret the internal information. Sometimes this external information may be of direct value in producing a forecast. An example of this would be when forecasts for a whole industry are used to provide basic information for forecasting the sales of a firm within that industry.

2.1.2 The second step — data reduction

The aim here is to pick out from the possibly vast amount of information obtained at the first step that which is regarded as most relevant. This is then reduced to a bare minimum. For example, in a sales forecasting situation the record of past sales for the firm would probably provide this minimum information. In deciding which items are to be included in this basic minimum information three criteria are frequently used. The minimum information should meet the requirements of

Relevance: Is the information the most directly relevant information available?

Reliability: How were the data obtained? How trustworthy are they?

Recency: Is it the latest information available? Will new information be available promptly?

When this minimum information has been thoroughly studied and its behaviour understood, we might start the reverse process of allowing other information to be used. For example, in sales forecasting we might use economic information or the assessments of the situation made by the sales force. However, for the moment we will consider only the use made of this minimum information. Though we have produced some criteria to be used in the selection of our information, we cannot usually do this in a purely objective fashion. Selection must be made on the basis of a broad and deep understanding of the practical situation. Later scientific study of this information will indicate whether our choice was good or bad. This will reveal itself by our success in obtaining good forecasts from the information. Unfortunately, we may not find out until too late that our information is not the best, and so our intuitive choice is a necessary and vital step in the process.

2.1.3 The third step — model construction

We now examine our minimum information with the object of identifying its structure and producing a precise specification for this structure. In most cases this specification will consist of a mathematical or statistical description of the structure; this description is what we normally call the model for the structure. To make this idea more concrete, let us consider again the data given in Table 1.1. In the previous chapter we sought to describe these data in purely verbal terms. Now we will describe them in mathematical terms. Let x_t denote the orders in week number t, $t = 1, 2, 3, \ldots$. For data (a) x_t is always 6,000 units, so we write

$$x_t = 6{,}000 \ (t = 1, 2, 3, \ldots)$$

For data (c) x_t becomes 9,000 in weeks 5, 10, 15, $\ldots$, and so we write

$$x_t = 6{,}000 \ (t = 1, 2, 3, 4, 5, 6, 7, 8, 9, 11, \ldots)$$

$$x_t = 9{,}000 \ (t = 5, 10, 15, 20, \ldots)$$

For data (d) we found that the orders differed from 6,000 by a value

14

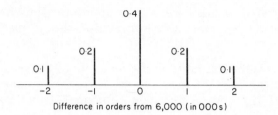

Difference in orders from 6,000 (in 000 s)

Figure 2.1   Probability distribution

that was not determinate but was governed by a statistical structure. Let us use the symbol ϵ_t to denote the chance or random difference from 6,000 that occurred in week t. This quantity ϵ_t can take on values ..., -2, -1, 0, 1, 2, ... (000s) with certain probabilities. The data given in Figure 1.1 gave the results of taking a sample of these random differences. Suppose the values given in Figure 2.1 gave the actual underlying probabilities of ϵ_t. The structure of the situation can thus be imagined as a vast hat containing tickets labelled $-2,000$, $-1,000$, 0, 1,000 and 2,000, in the proportions given in Figure 2.1, these proportions in fact being the probabilities of choosing the respective ticket numbers. At the end of week t a ticket is chosen at random, its number we call ϵ_t, and this number is added to 6,000 to give x_t. Thus the model is

$$x_t = 6,000 + \epsilon_t$$

where, to use more precise terms, ϵ_t is a random variable having the probability distribution given in Figure 2.1. For the data of line (e) of Table 1.1 a trend is added to the structure. This trend, as shown in Figure 1.2, is expressed mathematically as

Trend value at week t = 1,000t

Hence the orders are modelled by the equations

$$x_t = 6,000 + 1,000t + \epsilon_t$$

In our example we have taken the data as the starting point and then constructed the model. The illustration of the tickets in the hat suggests that it is always possible to start with a model and construct from it data having that structure. The purely mathematical part of the model presents little problem, but generating values for ϵ_t can get quite complicated. A literal hat would do the job for the distribution in Figure 2.1. For more complicated distributions we need more elaborate equipment, most usefully a computer. There are two reasons why it is useful to be able to generate forged data having the properties of any given model. The first is that it gives us experience of the behaviour to expect from data of the given structure, and, secondly, it enables us to examine the properties of different methods of forecasting. The experience gained of the behaviour of different kinds of model is a great help in deciding what model to consider for our own particular set of data.

If we used the probability distribution of Figure 2.1 to generate a set of ϵ_t to use in the trend model, we would obtain a set of data which was not identical to our original data in Table 1.1 but which would have the same structure. Thus we see that the model is not identical with the set of data, nor does it completely explain why the data behave in this or that detailed fashion. The model simply gives a mathematical or statistical formulation of the situation, a formula that can generate numbers showing a similar behaviour to the observed data.

2.1.4 The fourth step — model extrapolation

The important difference between the model and the data for our purposes is that the latest piece of information completes the data, whereas the model will produce values for any value of t we care to substitute. Thus in the expression

$$x_t = 6,000 + 1,000t + \epsilon_t$$

we can substitute any value for t and not just the values $t = 1$ to 12 which represented the original data. The evaluation of the mathematical part, $6,000 + 1,000t$, for differing values of t is a matter of simple substitution. The evaluation of the random part, ϵ_t, will depend on what we are seeking to do. If the aim is to generate a set of data with the given properties, a numerical value must be found for ϵ_t from the appropriate probability distribution. If the aim is to forecast, a reasonable procedure would be to replace the unknown future value of ϵ_t by its theoretical average value, which is usually zero. Thus a forecast of x_{14}, denoted by $\hat{x}_{14}$, would be

$$\hat{x}_{14} = 6,000 + 1,000 \times 14 + 0$$
$$= 20,000$$

since in the distribution of Figure 2.1 the theoretical average, or expectation, of ϵ_t is zero. We thus extrapolate our model, our structure, into the future, assuming that the situation is stable.

The basic steps that we have discussed are represented in Figure 2.2. There are many problems associated with the construction of models that have been ignored in the previous general discussion, in particular the problems of choosing the form of model and of choosing the numbers to use in it. For example, how did we know that in the above illustration a trend model should be used with a starting value of 6,000 and a constant rate of increase of 1,000 units per week. In a real

16

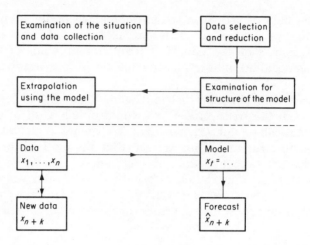

Figure 2.2   Four basic steps in statistical forecasting

situation we start with complete ignorance, i.e. with the model

$$x_t = ?$$

To decide what this question mark should be is called the problem of IDENTIFICATION, the problem of the choice of model. In the above data we have assumed that we have solved this problem and we have a model of the form

$$x_t = \alpha + \beta t + \epsilon_t$$

where α and β are constants, β being the slope of the line and α the intercept with the x-axis. We still, however, do not know what numerical values to put for the constant α and the slope β. The problem faced here is called the problem of ESTIMATION or model fitting, the choice of numerical values to use in the model. When we have solved this problem for the above model, we then might put

$$x_t = 6,000 + 1,000t + \epsilon_t$$

Notice that these are not once and for all problems to be dealt with before we start forecasting. At any stage our model will, we hope, be one that gives a good description of the data so far observed. However, as time progresses new data will be obtained and the structure may change or the increased amount of available information may reveal features that were not apparent before. There are two consequences of this possibility. The first is that the processes of identification and estimation need, in some fashion, to be continued as long as forecasting is continued. The second is that even if to date we have a very good model for the data which has given excellent forecasts, there is still no guarantee that this model will continue to give good forecasts in the future. The chances of a structure remaining stable decrease, the longer the period of time considered. It follows that the doubt about the reliability of our forecasts increases, the greater the period into the future over which we extrapolate our model. The possible failure of our basic underlying assumption that we have correctly identified a stable structure means that the forecaster can never quite throw out his crystal ball.

2.2 Basic models

It was seen in our previous discussion that the choice of model is a basic step in forecasting. A natural reaction to this by many practical people is to feel that since their situation is probably unique they will have to invent their own models to fit their situation. Since they may feel this to be beyond them, they might lose hope in statistical forecasting. Fortunately, it has been found in practice that a very wide range of data can be adequately modelled by a relatively small number of basic models. Table 2.1 gives a list of some of these basic models. Part II of this book considers the use of each of these models in forecasting. Often these models provide all that is needed for practical forecasting. Sometimes, however, they will need to be combined, modified and tailored in some fashion to the problem in hand. Some approaches to doing this and some examples are given in Part III of this book.

Let us use the models of Table 2.1 to give a first introduction to some terminology. Models such as (a), (b) and (c) in the table are perfect mathematical forms. For example, if we had at times 0, 1, 2, 3, 4, data 2, 4, 6, 8, 10 from the perfect linear trend, then any two of the observations could be used to show that $\alpha = 2$ and $\beta = 2$; so a perfect forecast of the next observation would be $12(= 2 + 2 \times 5)$. Such a model we will term a *deterministic* model. As we discovered in example 1.4, such models rarely describe the real world owing to the chance, or *random*, variation that is shown by most data. Random sequences such as that shown in (g) of Table 2.1 are often used to model this aspect. So, as we have seen, we might have data modelled by

x_t = linear trend + random sequence

or, mathematically,

$x_t = \alpha + \beta t + \epsilon_t$

where ϵ_t is used to denote a sequence of independent random quantities scattered above and below zero. As the deterministic part is a fixed basis for this model, we will still often refer to this type of model as essentially deterministic. Unfortunately, data from such a model will look like that shown in Figure 1.2, so it will not be possible to take just

Table 2.1 Some common models

Name	Plot	Form
(a) Constant mean		$x_t = \mu$ for all t
(b) Linear trend		$x_t = \alpha + \beta t$ for all t
(c) Linear regression		$x_t = \alpha + \beta z_t$ z_t being another variable
(d) Autoregression		$x_t = \alpha + \beta x_{t-1}$ x_{t-1} being the previous value of x
(e) Periodic		x_t repeats pattern every T observations $x_{t+nT} = x_t$ for all t $(n = 1, 2, 3)$
(f) Exponential growth		$x_t = \alpha\, e^{\beta t}$ for all t
(g) Random normal sequence		x_t is normally distributed with mean μ and variance σ^2 values of x_t independent of each other for all t

two observations to find α and β. We will need to develop methods to find values for α and β, to *estimate* α and β, so that the model we get shows a good fit to the data and thus can reasonably be extrapolated to obtain a forecast. The problems of estimating *parameters*, such as α and β, and thus *fitting* models to data will be basic to much of our discussion in the following chapters.

Let us look now at the model described by (d) of Table 2.1. As it stands, the situation described by this equation is deterministic. However, if we become more realistic and write

$$x_t = \alpha + \beta x_{t-1} + \epsilon_t$$

to allow for the chance variation of the real world, we obtain a very different situation. As x_t is influenced by the previous x, it must also be influenced by previous random variables ϵ. Thus x_t is influenced by the whole past history of the sequence of ϵs. Such a model is an example of a *stochastic* model. Such models form the base for a large number of forecasting methods. Many forecasting methods are based on either a deterministic or a stochastic model. We will tend to look at them separately, but clearly there are situations where one would seek to combine these types of model.

Models (b), (c) and (d) all have the form

$$x_t = \alpha + \beta(\quad)$$

The parameters thus occur singly in separate terms of the model, the terms being added together. The influence of the parameters and the effects of changing them can thus be considered independently for each parameter. Such models are said to be *linear* in the parameters. Problems of estimating parameters and forecasting tend to be simplest for linear models. Other models, such as the growth curve in (f) in Table 2.1, do not have this property. Here the model is *non-linear* in the parameters; the effect of changing β in (f) will depend on the value taken by α. In studying such growth curves we will be faced with more difficult problems of estimation and forecasting.

2.3 Models of global and local validity

When we use models such as those given in our list, we shall sometimes find that they are being used in a much more rough and ready fashion than is indicated by the mathematics. In particular, we shall often use them not in the belief that they exactly describe the underlying structure of the situation but in the faith that, at least in the recent past and, one hopes, the near future, they will give a reasonable description of this underlying situation. We shall thus consider two possible situations that could occur in practice. In the first, the structure is regarded as highly stable and the chosen model as the truth about the underlying structure of the data. The model for such a situation will be called a global model. In the second situation, the structure is believed to be stable in the short run but not necessarily in the long run. Slow changes in structure may occur in our data which will not affect us too adversely for forecasting over short periods of time. When examining forecasting at an advanced level, attempts are usually made to incorporate any slow changes in structure into a more complex model. It is, however, often sufficient to treat such data as coming from, what we will term, a model of local validity, or a local model.

There is no difference in the mathematical or statistical formulation of these two types of model, global and local. The difference lies in the way in which we make use of the models. Thus when we use a local constant mean model we shall forecast the future on the assumption that the mean in the near future will be the same as the current mean. We will not be willing, however, to assume that the current mean is the same as that in the past when we first obtained our data. We will, in future chapters, discuss models from two viewpoints: firstly, from the global viewpoint, with the model assumed to be a proper description of the underlying situation; secondly, from a local viewpoint, with the model regarded as an approximation valid only locally in time. There are a number of different situations that would lead to the use of models as approximations only. These are

(a) situations where there is not enough data available to enable us to study the underlying situation in detail;
(b) situations where for reasons of simplicity and economy we deliberately use a simpler model than the one that we think might be appropriate;
(c) situations where the quantity being forecast is known to be influenced by factors that are not stable but which change in time in a relatively slow fashion.

Experience suggests that using models as local approximations often leads to better forecasts than are obtained from their global use. That this is so may well relate to the fact that with data in time we are always effectively in situation (a). The underlying situation by its nature stretches from some past time to some unknown future time. Even if by magic we had data from all this time, it would only represent one observation on the entire process. In practice we only have data from a small section of time and thus, in effect, we have less than one observation. That we can get anything at all from this situation depends on major assumptions about the nature and stability of the process. Against such a background, using our models as approximations to sections of the underlying process seems a safer course than using our models as global descriptions of the situation.

2.4 Forecasting models and formulae

People who have studied statistics will be familiar with the ideas of statistical models for data, such as those just referred to. It is therefore important to draw a distinction between the use of models in forecasting and the more usual statistical uses. In fact, we might refer to the models in section 2.3 as forecasting models. Though they are statistical or mathematical models in form, their required relation to the data is different in forecasting from their relation in ordinary statistics. In statistical model building we are just as interested in how the model describes the first observations as in how it describes the most recent observations. In forecast model building it is the future that interests us, and as long as the model is good at giving forecasts we do not mind how badly it fits the data of the dim and distant past. Thus we will often feel justified in using a model that obviously does not fit the entire set of data. Provided it gives a good fit to the most recent data, it is a potentially good forecasting model. Putting this a different way, our attitude to models in forecasting means that we are as happy with a local model as with a model of global validity, as long as we get good forecasts.

Much of this book will be devoted to describing how to obtain formulae for forecasting future values in particular forecasting models. We will call such formulae *forecasting formulae*. For example, given a model of the form

$$x_t = \mu + \epsilon_t$$

where μ is a constant mean and ϵ_t is a purely random normal sequence, and given past data $x_1, x_2, \ldots, x_t$, we wish to forecast the future value x_{t+k}. Denote by $\tilde{x}_{t,k}$ the forecast made at time t of the value of x at k time units into the future, referred to as a forecast with lead time k. An appropriate forecasting formula for x_{t+k} is

$$\tilde{x}_{t,k} = \frac{1}{t}(x_1 + x_2 + \ldots + x_t)$$

Usually the forecasting formulae will be obtained from some specific forecasting model. However, it often occurs that a particular forecasting formula provides reasonable practical forecasts for a number of forecasting models. Hence we will sometimes be interested in fore-casting formulae in themselves without reference to forecasting models.

Another forecasting formula that might have been used in our example, had we felt that the model should be used as a local model rather than a global model, is

$$\tilde{x}_{t,k} = (1-a)(x_t + a x_{t-1} + a^2 x_{t-2} + \ldots)$$

where a is some number between 0 and 1 which is chosen to give the best forecast in some sense. This number is a parameter of the forecasting formula and will be referred to as a *forecasting parameter*, as distinct from a model parameter such as μ. The choice of such a forecasting parameter will control the properties of the forecasts. Thus, if in the example a is small, the formula will strongly emphasize the recent values, x_t, x_{t-1}, etc. If, however, a is chosen close to 1 and t is large, the formula gives much the same answer as that given by the first formula.

Chapter 3

Forecasting criteria

3.1 The need for criteria

Consider the following four problems:

Problem A

A sales manager collects together records of past monthly sales for a particular product. How does he now find a forecasting formula that gives good forecasts for this data?

Problem B

A member of his staff suggests a particular method of forecasting. How does the sales manager test this method to see if it is any good?

Problem C

Having examined several products, the sales manager decides that they all have a similar structure which could reasonably be modelled by a linear trend with an added random variation adequately described by a purely random normal sequence. How does he now find a forecasting formula that gives good forecasts for this particular model?

Problem D

His enthusiastic member of staff advocates his pet method as the best method for this type of model. How does the sales manager test the success of this method for his model?

The ability to find answers to questions such as these is obviously basic to the whole exercise of practical forecasting. However, in each problem it is assumed that we know what is meant by a *good* forecast.

We forecast sales of 6,539 items and, lo and behold, we sell 6,539 items — a perfect forecast! But suppose we sell 6,502 items; can we say that our forecast was good or not? Without more information we cannot answer this question. The error of 37 items may be large or small relative to an acceptable error in the given situation. We cannot say whether a forecast is good until we are agreed on what is meant by the term 'a good forecast'. The two words 'good' and 'forecast' are really inseparable A bad forecasting method is just not a tenable method of forecasting. The consequence of this situation is that we must postpone a discussion of how to forecast until the nature of a 'good forecast' has been investigated. The aim of this chapter is to carry out this investigation.

A careful examination of the four problems at the beginning of this section shows that the subject matter to which the criterion of goodness is applied may be different in different situations. In two of the problems the subject matter is the data itself; in the other two it is a theoretical model. Further, the objectives may be different in different situations. In two of the problems the object is to derive a forecasting formula; in the other two it is to test out a given formula. There are thus four classes of problem, as is shown in Table 3.1.

Table 3.1 A classification of problems needing forecasting criteria

Subject matter of criteria	Objective of use of criteria	
	Deriving a forecasting formula	Testing a forecasting formula
Data	A	B
Model	C	D

We found in Chapter 2 that obtaining a forecasting formula, given a set of data, involved first replacing the data by some appropriate model. Thus, obtaining a formula for a set of data usually reduces to the derivation of a formula for a model. The word 'usually' is inserted here since we will later look at methods where the model-building aspect is ignored. The 'best' formula is often found by testing out several forecasting methods using the data and choosing the best according to some criterion. In this case the problem of deriving a formula is replaced by the problem of testing a formula. Thus, in general, problem A is reduced, most frequently, to problem C, though sometimes to problem D.

The problem of deriving formulae for a given model to meet certain criteria can be a very difficult one. It will be dealt with at different levels at various parts of this book It is sufficient to note at this stage that, as is reasonable, the criteria that can be used as the basis for deriving a forecasting formula for a given model are usually the same as those which are used for testing out any given formula on the model. Exceptions to this rule sometimes occur, in particular a forecasting formula may be first derived according to one particular criteria and afterwards tested out against other criteria to provide additional information about its behaviour. However, for the purposes of this introductory chapter, we will put together problems of type C with those of type D under the latter heading.

The outcome of the above discussion is that a preliminary investigation of forecasting criteria may be made by considering only the case of the problem of testing a given forecasting formula. We consider in particular how to measure the success of a given forecasting formula (a) for a particular set of data and (b) for a particular model.

3.2 Classification of criteria for testing forecasting formulae on data

In this section we shall consider the situation where we have a set of data, $x_1, x_2, \ldots, x_n$, and a forecasting formula which can be used to obtain a forecast from such data. We will denote the forecast of x_t by $\hat{x}_t$, where $\hat{x}_t$, is obtained from the forecasting formula using some or all of the observations that existed before time t. The discussion of forecasting criteria is thus a discussion of the sense in which $\hat{x}_t$ might be close to x_t.

To make the discussion as concrete as possible, consider the data in Tables 3.2 and 3.3. It will be seen from Figure 3.1, which charts the data of Table 3.2, that the first part of the data is constant at 6 units, the second section shows random variation about this constant value and then, after the fifteenth observation, the underlying mean of the set of data jumps from 6 units to about 10 units. The behaviour of the data in Table 3.3 is similar to that of Table 3.2 for the first six observations, but then there is a steady upward trend. Columns $\hat{x}_A$ and $\hat{x}_B$ of these tables and the corresponding plots in Figures 3.1 and 3.2 show the forecasts for this data obtained by two different forecasting formulae, A and B. It is seen that both methods give perfect forecasts where the data is constant at the value of 6. Where the data show random variations, both forecasts also show random variations; however, the variability in the forecasts obtained by method A is much greater than that shown by the method B forecasts. Method B is said to 'smooth' the data more than method A. Where there is a step change in the underlying mean for the data, method A very quickly picks up the new mean and varies randomly about the new value. Method B is very slow to follow the data and even after some ten observations the forecast is still lagging considerably below the new mean value. Where

26

Table 3.2 Illustrative forecast results I

| Time | Data | Forecasts | | Errors | |
t	x	$\hat{x}_A$	$\hat{x}_B$	e_A	e_B
1	6	—	—		
2	6	6	6	0	0
3	6	6	6	0	0
4	6	6	6	0	0
5	6	6	6	0	0
6	6	6	6	0	0
7	7	6	6	1.00	1.00
8	5	6.5	6.10	−1.50	−1.10
9	4	6	5.99	−2.00	−1.99
10	8	4.5	5.79	3.50	2.21
11	6	6	6.01	0.00	−0.01
12	7	7	6.01	0.00	0.99
13	7	6.5	6.11	0.50	0.89
14	5	7	6.20	−2.00	−1.20
15	10	6	6.08	4.00	3.92
16	9	7.5	6.47	1.50	2.53
17	11	9.5	6.72	1.50	4.28
18	12	10	7.15	2.00	4.85
19	10	11.5	7.64	−1.50	2.36
20	9	11	7.87	−2.00	1.13
21	8	9.5	7.99	−1.50	0.01
22	10	8.5	7.99	1.50	2.01
23	11	9	8.19	2.00	2.81
24	10	10.5	8.47	−0.50	1.53
25	9	10.5	8.62	−1.50	0.38

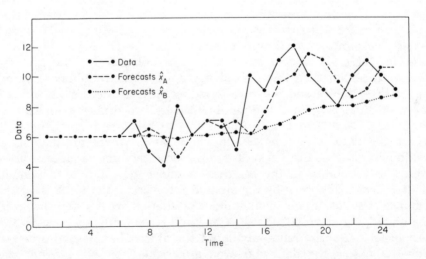

Figure 3.1 Plots of data and forecasts from Table 3.2

Table 3.3 Illustrative forecast results II

Time	Data	Forecasts		Errors		Transients	
t	x	$\hat{x}_A$	$\hat{x}_B$	e_A	e_B	z_A	z_B
1	6	—	—				
2	6	6	6	0	0	0	0
3	6	6	6	0	0	0	0
4	6	6	6	0	0	0	0
5	6	6	6	0	0	0	0
6	6	6	6	0	0	0	0
7	7	6	6	1.0	1.0	0.5	9.00
8	8	6.5	6.10	1.5	1.90	0	8.10
9	9	7.5	6.29	1.5	2.71	0	7.29
10	10	8.5	6.56	1.5	3.44	0	6.66
11	11	9.5	6.90	1.5	4.10	0	5.90
12	12	10.5	7.31	1.5	4.69	0	5.31
13	13	11.5	7.78	1.5	5.22	0	4.88
14	14	12.5	8.30	1.5	5.70	0	4.30
15	15	13.5	8.87	1.5	6.13	0	3.87
16	16	14.5	9.49	1.5	6.51	0	3.49
17	17	15.5	10.14	1.5	6.86	0	3.14
18	18	16.5	10.82	1.5	7.18	0	2.82
19	19	17.5	11.54	1.5	7.46	0	2.54
20	20	18.5	12.29	1.5	7.71	0	2.29
21	21	19.5	13.06	1.5	7.94	0	2.06
22	22	20.5	13.85	1.5	8.15	0	1.85
23	23	21.5	14.67	1.5	8.33	0	1.77
24	24	22.5	15.50	1.5	8.50	0	1.50
25	25	23.5	16.35	1.5	8.65	0	1.35
Large t	t	$t-1.5$	$t-10$	1.5	10.00	0	0.00

there is a perfect linear trend, the forecasts obtained by method A lag a fixed 1.5 units behind the observations. Method B also gives forecasts that lag behind the data, but it is much slower at 'realizing' that a change has taken place, and even after 20 time units the forecasts are not increasing at the same rate as the observations, though from Figure 3.2 it is seen that its rate of increase is slowly catching up with that of the data. When the rates of increase become the same, the forecast will lag behind the data by a constant 10 units.

Let us return now to the basic problem of what constitutes a criterion of success for a forecasting method when applied to a set of data. From the examples it is seen that there are various ways in which the forecast can differ from the data. There is thus not just one criterion of success in forecasting but many. Two of the great difficulties in practical forecasting are the choice of which criteria to use and the balancing of opposing criteria to get the best of several worlds. For example, we might require both that a forecast does not pick up

28

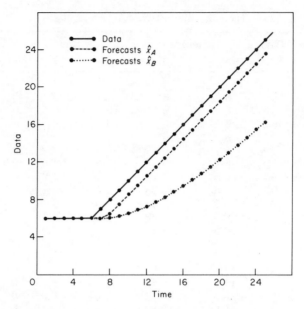

Figure 3.2   Plots of data and forecasts from Table 3.3

too much of the random variation in the data, as method A does, nor that it lags too far behind any new feature, as method B does. When dealing with testing out forecasting formulae, both these aspects must be studied. It must be recognized, however, that if a method is to be sensitive enough to adjust quickly to major changes in the structure of the data, it will also be sensitive to the random variations in the data. Conversely, if we wish to reduce the response to random variation, giving forecasts that are 'smoother' than the data, we will also prevent the forecast from responding quickly to changes in the structure of the data.

It is of value to classify the features that interest us when trying out forecasting formulae on data (and also on models). Three main headings can be used. These are as follows.

(a) Statistical features

This heading includes the amount of smoothing produced by the formula and the statistical distribution of the forecast errors.

(b) Transient features

These are the features that show up when a change occurs in the underlying structure but which in time die away, e.g. the differing rates of response to the new trend in Table 3.3 and Figure 3.2.

(c) Steady-state features

These are the features shown by forecasts when used in a stable situation. An example of these is the constant lag shown by both methods in Table 3.3 after the transient effects have died away.

A detailed study of these features is usually only possible when testing forecasting methods on models. When considering a set of data, lack of knowledge of the underlying structure of the data makes detailed study under these headings difficult and tends to blur the distinction between them. However, they are still of sufficient use to justify their separate consideration in more detail.

3.3 Statistical features

To study the elementary statistical features of a situation, it must be assumed that the situation has a stable structure. For this to be the case when forecasting the data must have a stable structure and hence there should be no transient features in the situation. If there are clear changes in structure in the data, the study of statistical properties should be based only on data sufficiently far from the changes that the affects of transient features can be neglected.

To illustrate the statistical features, consider observations 7 to 14 in Table 3.2. These show a simple random variation about a constant mean. Table 3.4 repeats this data for ease of discussion. The fourth and fifth columns of this table show the forecast errors e_t, which we will always take as being actual x_t minus forecast $\hat{x}_t$,

$$e_t = x_t - \hat{x}_t$$

Thus a positive error means that the value that occurs is larger than the forecast value. It should be noted that some writers define the error the other way round, as $\hat{x}_t - x_t$. An examination of the errors in Table 3.4 shows that the errors from method A show a greater spread than those of method B, but the average error is smaller. If we examine the average error in the data of Table 3.2 for the trending region, this average will be positive, 1.5 for method A, indicating the lag of the forecast behind the trend. Hence a study of the errors requires at least a study of (a) their average (mean) value and (b) their spread. The mean error $\bar{e}$ for a set of errors $e_1, e_2, \ldots, e_n$ is defined simply as

$$\bar{e} = \frac{1}{n}(e_1 + e_2 + \ldots + e_n)$$

$$= \frac{1}{n} \sum_{i=1}^{n} e_i \text{ (using the summation notation)}$$

A good method of forecasting will usually be required to give a small

Table 3.4 Statistical analysis of forecasts

(a)

Time t	Data x	Forecasts $\hat{x}_A$	Forecasts $\hat{x}_B$	Errors e_A	Errors e_B	Absolute errors $\lvert e_A \rvert$	Absolute errors $\lvert e_B \rvert$	Squared errors e_A^2	Squared errors e_B^2
7	7	6	6	1.0	1.00	1.0	1.00	1.00	1.00
8	5	6.5	6.10	−1.5	−1.10	1.5	1.10	2.25	1.21
9	4	6	5.99	−2.0	−1.99	2.0	1.99	4.00	3.96
10	8	4.5	5.79	3.5	2.21	3.5	2.21	12.25	4.88
11	6	6	6.01	0.0	−0.01	0.0	0.01	0.00	0.00
12	7	7	6.01	0.0	0.99	0.0	0.99	0.00	0.98
13	7	6.5	6.11	0.5	0.89	0.5	0.89	0.25	0.79
14	5	7	6.20	−2.0	−1.20	2.0	1.20	4.00	1.44
			Totals	−0.5	0.79	10.5	9.39	23.75	14.26
			Means	−0.062	0.099	1.31	1.17	2.97	1.78
			Root mean squares (RMS)					1.72	1.31

(b) Summary

	Method A	Method B
Mean error	−0.06	0.10
Mean absolute error (MAE)	1.31	1.17
Mean square error (MSE)	2.97	1.17
Root mean square error (RMSE)	1.72	1.31

mean error. Where this average error becomes large, perhaps due to failure to follow accurately some systematic structure in the data, the forecasting formula becomes suspect and needs adjusting or replacing by another which is suitable for this particular type of structure. A systematic deviation of the forecast errors from zero is referred to as a 'bias' in the forecasts. The mean on its own is not enough, since it may well be close to zero while the actual errors are very large, positive and negative errors tending to cancel each other out. To eliminate the sign for the purpose of measuring spread, one may either just ignore it and consider the absolute errors, denoted by $\lvert e_t \rvert$, or remove it by squaring. The first approach measures spread by the mean absolute error (MAE),

$$\text{MAE} = \frac{1}{n} \sum_{i=1}^{n} \lvert e_i \rvert$$

sometimes also called the mean deviation of error. The second approach uses the mean square error (MSE),

$$\text{MSE} = \frac{1}{n} \sum_{i=1}^{n} e_i^2$$

Table 3.4 gives the calculation of these quantities. It is seen that both quantities indicate quite clearly that method B gives errors having a smaller spread than method A.

To make the units of the mean square error the same as those of the mean error and the mean absolute error, its square root must be taken to give the root mean square error (RMSE). The choice of whether the mean absolute error or the root mean square error is used is based on practical considerations of the ease of calculation and the use to be made of the calculated measures. It should also be noted that the root mean square error places greater emphasis on the large errors than does the mean absolute error.

If a bias is found to exist, it may be advisable to measure the spread of the errors about their average rather than about the zero value. Thus the mean square error would be replaced by the 'sample variance' defined by

$$s^2 = \frac{1}{n} \sum_{i=1}^{n} (e_i - \bar{e})^2$$

A little calculation will show that this is related to $\bar{e}$ and the mean square error by

$$s^2 = \text{MSE} - \bar{e}^2$$

The square root of this is called the sample standard deviation, s.

In many of the computer programmes available commercially and in much of the literature on forecasting, the root mean square error is used as the main criteria of forecast success. It is seen that

$$\text{MSE} = \bar{e}^2 + s^2$$

Thus the mean square error is the sum of two contributory factors, one measuring any bias there might be in the errors and the other the variability of the errors about the mean error. An examination of the mean square error without taking $\bar{e}$ into account may thus be misleading or at least not as informative as a separate study of $\bar{e}$ and s^2 or, equivalently, $\bar{e}$ and MSE.

In the same way that the mean square error can be modified to give s^2, we may modify the mean absolute error by replacing the errors by their deviations from the mean error, to obtain the mean absolute deviation of errors (MADE),

$$\text{MADE} = \frac{1}{n} \sum_{i=1}^{n} |e_i - \bar{e}|$$

We are now in a position to say that from a statistical point of view the simplest criteria of good forecasts are the requirements of small average error and small mean square error or mean absolute error. These are sometimes sufficient in terms of forecasting criteria, but in testing out a forecasting formula on any reasonably extensive set of data it is advisable to study the forecast errors in some more detail. This is of value in giving a better 'feel' for the methods and data that are under study, and also possibly leads to ways of improving forecasts.

3.4 Transient features

Transient features only become apparent where the data exhibit a sudden change, a fracture, in their structure. The only way of examining such features in data is to attempt to identify the occurrence of such fractures and examine the behaviour of the forecast errors in these regions. With clear fractures and well-behaved data, as in Table 3.3 and Figure 3.2, the nature of the transients is clear. Table 3.3 exhibits the transient phenomena by simply removing the steady-state values from the errors. With the usual messy data of the real world such a procedure is difficult, if not impossible. Hence a study of the transient features is best done using models of the types of structure and fractures that either are shown by the actual data or might reasonably be shown by that type of data. The examples and discussion of section 3.2 show that in examining the success of a forecasting formula the existence of large transient features in the forecast errors is highly undesirable. Any effects that do occur should die away as rapidly as possible. Thus from Table 3.3 this criteria is satisfied far better by method A than by method B.

A type of transient feature of some importance in forecasting is due not to a change in structure but to the way in which the forecasting method is initiated. For example, some methods of forecasting require an initial forecast to be made by the user before the forecasting process is started on the actual data. If this initial forecast is a poor one, it will influence the forecasts for some time to come.

3.5 Steady-state features

In analysing steady-state features in data, we meet very similar problems to those in the previous section. One can examine the data to find regions that appear to be far removed from any fractures in the structure. This could avoid the problem of transient features. These regions can then be studied to see if there is any clear steady-state behaviour. The simplest way of doing this is to examine the average error over these regions. As was seen in the example of Table 3.3, a steady-state effect, such as lag, usually shows itself as a significant deviation of the average error from zero. There are, however, other types of steady-state behaviour for which the average error will be close to zero. In particular, if an oscillation occurs in the data which is followed by the forecasting formula with some lag, positive and negative errors will occur in about equal proportions and thus a small average error will occur; none the less, there is in this situation a clear steady-state feature. The best and simplest method of examining such features is to plot a chart of the errors against time. Chapter 15 gives examples of such plots.

When dealing with a set of data we are very limited with what can be done in investigating steady-state features, and hence again we must turn to the use of models. Here we would use both models chosen to describe the existing data and also models that describe features that could well occur in the future structure. The analysis of the steady-state effects, when the forecasting formula is applied to a model, is thus the crucial test of the formula. The existence of steady-state errors is obviously a disadvantage to any forecasting formula and it is necessary to design formulae so that they are unlikely to occur. Where their existence is acceptable, as a sacrifice made in order to obtain other advantages, the effects should be kept as small as possible.

3.6 Criteria for testing forecasting formulae on models

In section 3.3 we based our criteria for forecast success on the actual errors between our forecasts and our observations. From these errors we calculated such things as the average error and the mean square error. When we turn now to criteria as they apply to models, we have to replace these numerical quantities by corresponding theoretical quantities. To do this we make use of the statistical ideas of the expectation and the variance of the distribution of errors. Table 3.5 summarizes the definitions and formulae needed. As can be seen from the table, all the quantities that we used in looking at the statistics of errors in data have corresponding quantities for a population of errors arising from a model. For the detailed theory of the expectation (sometimes called the population mean) and variance the reader is referred to any standard statistical text (e.g. see Hogg and Craig, 1970; Mood and Graybill, 1963). We will limit ourselves to illustrating the ideas with some forecasting examples.

Let us first apply the idea of expectation to a study of the transient and steady-state errors that occur when a forecasting formula is used with a particular model. For the sake of illustration we will use a model which consists of a linear trend with an added random variation. Thus we have for our observations x_t,

$$x_t = \alpha + \beta t + \epsilon_t$$

where the constant α is the mean of the trend at time zero, β is the

34

Table 3.5 Expectation and related ideas

Distribution of errors	$P(e)$ = probability (error = e), for discrete errors or $f(e)$ = probability distribution function of errors, for continuous case
Expectation of error (bias)	$E(e) = \sum\limits_{\text{all } e} eP(e)$, for the discrete case or $E(e) = \int\limits_{\text{all } e} ef(e)de$, for the continuous case
Expectation rules	$E(ae + b) = aE(e) + b$, for constants a and b $E(\sum\limits_i e_i) = \sum\limits_i E(e_i)$
Mean square error for distribution	$MSE = E(e^2)$
Mean absolute error	$MAE = E(\lvert e \rvert)$
Variance of error	$\text{Var}(e) = E[\{e - E(e)\}^2]$
Variance rules	$\text{Var}(e) = MSE - E(e)^2$ $\qquad\quad = MSE - \text{bias}^2$ $\text{Var}(ae + b) = a^2 \text{Var}(e)$ $\text{Var}(\sum\limits_i e_i) = \sum\limits_i \text{Var}(e_i)$, only if the errors e_i are independent

slope and the random variables ϵ_t are independent observations on a population with a normal distribution, mean zero, i.e. $E(\epsilon_t) = 0$, and variance σ^2. Another way of describing this model would be to say that the x_t come from a normal population with mean $\alpha + \beta t$ and variance σ^2. However, the above form, in which the deterministic and random parts of the situation are clearly separated, is most useful for the type of study we are undertaking. As a forecasting formula we will use the forecast $\hat{x}_{t+1}$ of x_{t+1} given by

$$\hat{x}_{t+1} = x_t$$

i.e. we simply use the last observation to forecast the next. This is the simplest formula to use here for illustrative purposes. It corresponds to the intuitive forecasts made by the type of individual who uses only his own experience and feel for the situation as the basis for forecasting but unfortunately has only a short memory. Having stated the model and the forecasting formula, we need only to bring the two together to obtain an expression for the forecast error. Thus,

$$e_{t+1} = x_{t+1} - \hat{x}_{t+1}$$
$$= x_{t+1} - x_t$$
$$= \{\alpha + \beta(t + 1) + \epsilon_{t+1}\} - \{\alpha + \beta t + \epsilon_t\}$$

Hence,

$$e_{t+1} = \beta + \epsilon_{t+1} - \epsilon_t$$

Using the properties of expectations given above, the expectation, which is the average of the population of the errors, is

$$
\begin{aligned}
E\,(e_{t+1}) &= E\,(\beta + \epsilon_{t+1} - \epsilon_t) \\
&= E\,(\beta) + E\,(\epsilon_{t+1}) - E\,(\epsilon_t) \\
&= \beta
\end{aligned}
$$

since the $E(\epsilon_t) = 0$ for all t from the definition. Thus the forecast is biased in the sense that the population average of the errors is β, so that on average the forecasts lag behind the observations by an amount β. This is a steady-state lag which remains the same over all time. To illustrate a transient phenomena, consider what happens if at some time T there is a sudden change in the situation which makes α become $\alpha + \delta$, this new value becoming permanent after time T. Since we have shown that e_t does not depend on the value of α, the expression we obtained for e_t will remain unaltered for times before and after e_T. The value of e_T will, however, be different and we will have

$$
\begin{aligned}
e_T &= x_T - \hat{x}_T \\
&= x_T - x_{T-1} \\
&= \{\alpha + \delta + \beta T + \epsilon_T\} - \{\alpha + \beta\,(T-1) + \epsilon_{T-1}\} \\
&= \beta + \delta + \epsilon_T - \epsilon_{T-1}
\end{aligned}
$$

Thus the bias is as before, except at time T when the change in α produces a transient effect in the form of sudden increase by δ which dies away immediately.

Let us now look at the spread of the errors. As we know the errors to be biased, it is more useful to look at the variance than the mean square error. This variance is the population quantity, from the model, which corresponds to the sample variance s^2, discussed in section 3.3. From the above results we now have

$$
\begin{aligned}
\mathrm{Var}(e_{t+1}) &= \mathrm{Var}(\beta + \epsilon_{t+1} - \epsilon_t) \\
&= \mathrm{Var}(\epsilon_{t+1}) + \mathrm{Var}(\epsilon_t) \\
&= 2\sigma^2
\end{aligned}
$$

using the fact that ϵ_t and ϵ_{t+1} are independent with variance σ^2, and letting $b = \beta$ and $a = -1$ in the formula for $\mathrm{Var}(a\epsilon_t + b)$. We see then that the variance of the forecast errors is twice the variance of the individual observations. This is a large error variance and suggests that our forecasting formula would not be of much practical use for data from this model. The mean square forecasting error is

$$E(e^2) = 2\sigma^2 + \beta^2$$

36

from the relation in Table 3.5 between the mean square error, variance and bias.

We have now seen that by substitution of the model in the forecasting formula we may investigate the main steady-state, transient and statistical properties of a forecasting situation.

3.7 Study by simulation

The forecaster can investigate quite complicated formulae and models using the ideas illustrated in the previous section. However, it frequently occurs that formulae and models get too complicated for this type of investigation or alternatively some information may be required without going to all this mathematical trouble. One method of study that can universally be applied is a study by simulation. To carry out a simulation study, artificial sets of data are constructed using the model, and the forecasting formula is then applied to these data. The results of the forecasts are then analysed by the usual methods of studying data discussed in sections 3.3, 3.4 and 3.5. For fairly simple cases this can be done by hand quite easily. If random models are to be studied in detail or a large number of possibilities are to be looked at, then computer simulation becomes advisable.

The data in Tables 3.2 and 3.3 are simple examples of data forged to show certain properties. Table 3.6 shows a working layout for

Table 3.6 Working layout for simulating a linear trend

$x_t = \alpha + \beta t + \epsilon_t$ $\alpha = 50$
$\beta = 2$
$\sigma = 2$, the standard deviation required
$\eta_t \cap N(0, 1)$ obtained from tables of random normal deviates

t	α	βt	η_t	$\epsilon_t = \sigma\eta_t$	$x_t = \alpha + \beta t + \epsilon_t$
0	50	0	−1.085	−2.170	47.830
1	50	2	−0.277	−0.554	51.446
2	50	4	−0.251	−0.502	53.498
3	50	6	0.817	1.634	57.634
4	50	8	−0.123	−0.246	57.754
5	50	10	−0.002	−0.004	59.996
6	50	12	−1.007	−2.014	59.986
7	50	14	−2.226	−4.452	59.548
8	50	16	−1.580	−3.160	62.840
9	50	18	1.468	2.936	70.936
10	50	20	0.169	0.338	70.338
11	50	22	1.530	3.060	75.060
12	50	24	0.598	1.196	75.196

simulating data from a trend model. The table shows one run of the simulation using one set of ϵs. Sets of random normal deviates, as such quantities from a normal distribution $N(0,1)$ are called, may be obtained from tables (e.g. Rand Corporation, 1955) or generated by standard instructions on most computers. The simulation of data with a seasonal variation was illustrated in Table 1.3.

References

Hogg, R. V. and Craig, A. T. (1970). *Introduction to Mathematical Statistics.* Macmillan, New York.

Mood, A. M., and Graybill, F. A. (1963). *Introduction to the Theory of Statistics.* McGraw-Hill, New York.

Rand Corporation (1955). *A Million Random Digits with 100,000 Normal Deviates.* The Free Press, Glencoe, Illinois.

Part II

Forecasting for some basic models

In the next eight chapters we will examine a range of models that form
the basis of most practical forecasting. The models discussed are

For each class of model we will examine various ways of obtaining
forecasts. In so doing we will develop a range of methods, principles
and approaches that have wide applicability in all types of forecasting;
these are discussed more generally in Chapter 12.

Chapter 4
The constant mean model

4.1 The global constant mean model

The form of this model is

$$x_t = \mu + \epsilon_t \ (t = 1, 2, \ldots)$$

where μ is the constant mean and ϵ_t is one of a sequence of independent random variables with zero expectation and constant variance σ^2. Table 1.1(d) showed some data from such a model which is characterized by a random variation about the constant level μ. We assume that observations $x_1, x_2, \ldots, x_t$ have been obtained and a forecast of a future observation x_{t+k} is required. We shall denote the forecast by $\tilde{x}_{t,k}$; the first subscript, t, denotes the time at which the forecast is obtained and the second subscript, k, denotes the time into the future which we span with our forecast, the lead time. The actual future value x_{t+k} can be expressed in terms of our model by

$$x_{t+k} = \mu + \epsilon_{t+k}$$

Thus to forecast x_{t+k} we shall have to assign values to both μ and ϵ_{t+k}. The random variable ϵ_{t+k} is by definition independent of the available information contained in $x_1, \ldots, x_t$ and hence we cannot give a forecast of its future value, save to note that its expected value will be zero. So we simply forecast ϵ_{t+k} by the value zero. The quantity μ is, for our global model, a constant over all time, and therefore an estimate of μ at the present time will also estimate its future value. The most natural estimate of the mean μ is the sample mean (average) based on all available data. Combining this with the zero forecast value for ϵ_{t+k} gives as our forecast the expression

$$\tilde{x}_{t,k} = (x_1 + x_2 + \ldots x_t)/t$$

We shall term this forecasting formula the 'explicit' form of our forecast, since it explicitly includes all our data.

The right-hand side of the above expression does not depend on k

41

because a constant mean is assumed and our observations are independent of each other. For the rest of this section we will drop the subscript k. It can be shown that, under certain conditions, the above forecast minimizes the mean square forecasting error, that is to say

$$E(x_{t+k} - \tilde{x}_t)^2$$

is minimum for this particular forecasting formula. As observations are usually obtained on a continuing basis, it is useful to try to put the forecasting formula in what is called recurrence form, i.e. in a form in which we can obtain the new forecast from the old forecast and the newly obtained data with the minimum of calculation. The new forecast $\tilde{x}_{t+}$, which will use the new observation x_{t+1}, can be related to the previous forecast in a quite simple fashion. Thus

$$\tilde{x}_{t+1} = [(x_1 + x_2 + \ldots + x_t) + x_{t+1}]/(t+1)$$

and hence

$$\tilde{x}_{t+1} = [t\tilde{x}_t + x_{t+1}]/(t+1)$$

Thus instead of having to keep a record of all past data, we need keep only the value $\tilde{x}_t$. When our next observation x_{t+1} is obtained, the above equation provides a simple way combining it with the last forecast $\tilde{x}_t$ to obtain the next forecast $\tilde{x}_{t+1}$. Table 4.1 gives a simple illustration of such a calculation, re-expressing the formula to give $\tilde{x}_t$ in terms of $\tilde{x}_{t-1}$ and $\tilde{x}_t$.

If we regard $\tilde{x}_t$ as the forecast of the next observation, x_{t+1}, then we can express the forecasting formula in yet another form. The basis of this third form is the use of the forecast error. Denote the forecast error at time t by e_t, where

$$e_t = x_t - \tilde{x}_{t-1}$$

Table 4.1 Forecasts for the global constant mean model — recurrence form

$$\tilde{x}_t = \frac{t-1}{t}\tilde{x}_{t-1} + \frac{1}{t}x_t, \quad \tilde{x}_0 = 0$$

t	x_t	x_t/t	$\tilde{x}_{t-1}$	$\frac{t-1}{t}\tilde{x}_{t-1}$	$\tilde{x}_t$
1	7	7.000	0	0	7.000
2	3	1.500	7.000	3.500	5.000
3	9	3.000	5.000	3.333	6.333
4	6	1.500	6.333	4.750	6.250
5	8	1.600	6.250	5.000	6.600
6	7	1.167	6.600	5.500	6.667

Table 4.2 Forecasts for the global constant mean model — error correction form

$$\tilde{x}_t = \tilde{x}_{t-1} + e_t/t, \quad e_t = x_t - \tilde{x}_{t-1}, \quad \tilde{x}_0 = 0$$

t	x_t	x_{t-1}	e_t	e_t/t	$\tilde{x}_t$
1	7	0	7.000	7.000	7.000
2	3	7.000	−4.000	−2.000	5.000
3	9	5.000	4.000	1.333	6.333
4	6	6.333	−0.333	−0.083	6.250
5	8	6.250	1.750	0.350	6.600
6	7	6.600	0.400	0.067	6.667

Then substituting for x_{t+1} in the last formula gives

$$\tilde{x}_{t+1} = (t\tilde{x}_t + e_{t+1} + \tilde{x}_t)/(t+1)$$

Hence,

$$\tilde{x}_{t+1} = \tilde{x}_t + e_{t+1}/(t+1)$$

Our new forecast is obtained from the old simply by adding a multiple of the last forecast error. A simple example is given in Table 4.2. The multiplying factor is 1/number of observations, which gets smaller and smaller as t increases. Thus the forecast gets less and less sensitive to forecast errors. A large change in structure could produce large errors, but yet because of the size of the divisor the forecast would be very little changed. The last formula is called the error correction form.

In the above discussion we have expressed our forecasting formula in three forms:

(a) the explicit form;
(b) the recurrence form;
(c) the error correction form.

We shall see in following chapters that these three forms not only represent different ways in which we can express the same forecasting formula but can also provide us with completely distinct approaches to the development of forecasting methods.

Having discussed different forms of the forecasting formula, we must next examine its properties. To do this, we simply make use of the model and of the simple rules that were given in Chapter 3.

Using the model we can re-express the forecast as

$$\tilde{x}_t = \sum_{t=1}^{t} x_i/t$$

$$= \sum_{t=1}^{t} (\mu + \epsilon_i)/t$$

44

Hence,

$$\tilde{x}_t = \mu + \sum_{t=1}^{t} \epsilon_i/t$$

Taking expectations of both sides and using the facts that $E(\mu) = \mu$ and $E(\epsilon_i) = 0$, we have

$$E(\tilde{x}_t) = \mu$$

Thus the forecast is unbiased. The forecast error is

$$e_{t,k} = x_{t+k} - \tilde{x}_{t,k}$$

using an obvious notation, so

$$e_{t,k} = \mu + \epsilon_{t+k} - \mu - \sum_{i=1}^{t} \epsilon_i/t$$

$$= \epsilon_{t+k} - \sum_{i=1}^{t} \epsilon_i/t$$

As $E(\epsilon_{t,k}) = 0$, for all k, the forecast is thus again proved unbiased for all future values.

Finding the variance of both sides of the above expression for $\tilde{x}_t$ and using the facts that $\text{Var}(\mu) = 0$ and $\text{Var}(\epsilon_i) = \sigma^2$, for all i, together with the independence of the ϵ values, gives

$$\text{Var}(\tilde{x}_t) = \text{Var}\left(\sum_{t=1}^{t} \epsilon_i/t \right)$$

$$= \sum_{i=1}^{t} \text{Var}(\epsilon_i)/t^2$$

Hence,

$$\text{Var}(\tilde{x}_t) = \sigma^2/t$$

Since the mean error is zero, as the forecast is unbiased, it follows that the mean square error is the same as the variance of error. Hence,

$$\text{MSE} = \text{Var}(x_{t+k} - \tilde{x}_t)$$

But x_{t+k} and $\tilde{x}_t$ are independent of each other and so

$$\text{MSE} = \text{Var}(x_{t+k}) + \text{Var}(\tilde{x}_t)$$

$$= \sigma^2 + \sigma^2/t$$

$$= \frac{t+1}{t} \sigma^2$$

Thus as t gets large the forecast estimates μ with high precision and the mean square error approaches σ^2, which is due simply to our inability to forecast the random variable ϵ_{t+k}.

Knowledge of the fact that the forecast is unbiased with error variance (mean square error,)

$$\sigma_e^2 = \frac{t+1}{t}\sigma^2$$

enables us to make much more useful forecast than the bare statement that the forecast is such and such a number. In particular, we can make statements about the probable range of values to be taken by the future observation. The most common form for such statements is called a confidence interval. We will outline the basic idea, but for details the reader is referred to any standard statistical text, e.g. Hogg and Craig (1970) and Mood and Graybill (1963).

Let us start by assuming that the forecast errors are normally distributed with zero mean and variance σ_e^2. It follows from the properties of the normal distribution that the probability of the forecast error lying in the interval $-1.96\,\sigma_e$ and $+1.96\,\sigma_e$ is 0.95. In general, the probabilities associated with ranges of values of a random variable are given by the areas under their distribution curves. In the normal error distribution (see Figure 4.1) these are determined by the spread of the curve as measured by σ_e. If we require different probabilities, then the 1.96 is replaced by some other appropriate

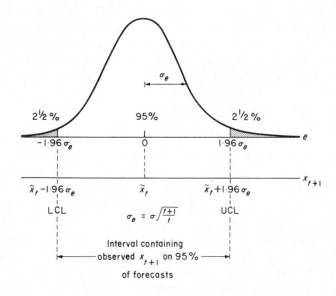

Figure 4.1 Prediction intervals

number obtained from tables of the normal distribution (e.g. see Fisher and Yates, 1938). We may write our statement as

$$\text{Prob} (-1.96\, \sigma_e < e < 1.96\, \sigma_e) = 0.95$$

Since

$$e_{t+1} = x_{t+1} - \tilde{x}_t \quad \text{and} \quad \sigma_e^2 = \frac{t+1}{t}\sigma^2$$

this may be simply rewritten as

$$\text{Prob}\left(-1.96\,\sqrt{\frac{t+1}{t}}\,\sigma \leqslant x_{t+1} - \tilde{x}_t \leqslant 1.96\,\sqrt{\frac{t+1}{t}}\,\sigma\right) = 0.95$$

and rewritten again as

$$\text{Prob}\left(\tilde{x}_t - 1.96\,\sqrt{\frac{t+1}{t}}\,\sigma \leqslant x_{t+1} \leqslant \tilde{x}_t + 1.96\,\sqrt{\frac{t+1}{t}}\,\sigma\right) = 0.95$$

Notice that, assuming σ is known, the terms forming the limits of the inequality are all known and can be calculated as two numbers. These we call the lower confidence limit (LCL) and the upper confidence limit (ULC). Thus,

$$\text{Prob} (\text{LCL} \leqslant x_{t+1} \leqslant \text{UCL}) = 0.95$$

The nature of this change from e to x_{t+1} is indicated in Figure 4.1. Care must be taken over the interpretation of such probability statements. The probability does not refer to the value of x_{t+1}, for this particular set of data, but rather it indicates that, in the long long run, using this method repeatedly will yield values of LCL and UCL which on 95 per cent of occasions will lie to either side of the observation being forecast. By way of example, suppose we have eight weeks' data believed to come from our model with $\sigma = 10$ and the forecast $\tilde{x}_t$ is 245 items. The two confidence limits are

$$\text{LCL} = 245 - 1.96\,\sqrt{\frac{9}{8}} \times 10 = 224.2$$

$$\text{UCL} = 245 + 1.96\,\sqrt{\frac{9}{8}} \times 10 = 265.8$$

Thus the 95 per cent confidence limits for the forecast are 224.2 and 265.8. The interval 224.2 to 265.8 is called the 95 per cent. confidence interval, or prediction interval.

Notice that, though for ease or notation we have talked about forecasting x_{t+1}, we could have talked of forecasting x_{t+k} without any change in the calculations. Similarly, we have treated the observations $x_1, \ldots, x_t$ as though they were made at equally spaced times called 1, 2, $\ldots$, t. The observations could have been made at any times, say T_1,

$T_2, \ldots, T_t$, without altering our results. The reason for this is that in this model the basis of the forecast is the estimate of μ which is assumed constant for all time. Thus it does not matter when the forecasts are made for or when the observations were taken. In all cases we get the same results and, in particular, the same confidence interval. We will see in later sections that this is not true in general.

In the previous discussion it has been assumed that σ is known, which is obviously very rarely true. When any parameter is not known, it must be estimated from the observations. To estimate σ^2, the variance of x, we use a simple modification of the sample variance of the data, which was defined in section 3.2. Here we define our estimate as

$$\hat{\sigma}^2 = \frac{1}{(t-1)} \sum_{i=1}^{t} (x_i - \bar{x})^2$$

where $\bar{x}$ is the mean of the available data, which is $\tilde{x}_t$ for the model of this section. If this estimate is used in place of σ^2 in the calculation of confidence limits, the value 1.96, or whatever was used from the tables, has to be replaced by the corresponding value from the table of the t-distribution. The reason is that when the constant σ^2 is replaced by the random variable $\hat{\sigma}^2$, which depends on the data, the normal distribution form of Figure 4.1 is no longer applicable and a distribution called a t-distribution is used instead (e.g. Fisher and Yates, 1938). If in our example the value of $\hat{\sigma}^2$ was 100 and the value of $t-1$, called the degrees of freedom, was 7, then looking up the 95 per cent value for 7 degrees of freedom in tables of the t-distribution gives the number 2.365 in place of 1.96. Recalculating the confidence limits gives

$$\text{LCL} = 245 - 2.365 \sqrt{\frac{9}{8}} \times 10 = 219.9$$

$$\text{UCL} = 245 + 2.365 \sqrt{\frac{9}{8}} \times 10 = 270.1$$

Thus the prediction interval is wider, which simply reflects the fact that, when σ is not known, we have less knowledge of the situation and cannot give as short an interval as before. In the above calculation we have obtained the error standard deviation σ_e from that of the data σ. As we will usually measure the errors directly it is more natural, and reliable, to estimate σ_e directly from the observed errors. We would thus use

$$\hat{\sigma}_e^2 = \frac{1}{t} \sum_{i=1}^{t} e_i^2$$

Table 4.3 Global constant mean model

Model	$x_t = \mu + \epsilon_t, t = 0, 1, \ldots, \epsilon_t$ independent, $E(\epsilon_t) = 0$, $\text{Var}(\epsilon_t) = \sigma^2$
Data	$x_t, \ldots, x_t$
Minimum MSE forecast of	x_{t+k} is $\tilde{x}_{t,k} = \tilde{x}_t$ independent of lead time k
Explicit form	$\tilde{x}_t = \sum_1^t x_t/t$
Recurrence form	$\tilde{x}_t = \tilde{x}_{t-1}(t-1)/t + x_t/t$
Error correction form	$\tilde{x}_t = \tilde{x}_{t-1} + e_t/t, e_t = x_t - \tilde{x}_{t-1}$
Forecast bias	0
Forecast MSE	$(t+1)\sigma^2/t$
95% prediction interval, σ^2 known	$\left(\tilde{x}_t - 1.96\,\sigma\sqrt{\frac{t+1}{t}}, \tilde{x}_t + 1.96\,\sigma\sqrt{\frac{t+1}{t}} \right)$

The prediction interval would then be between $\tilde{x} - c_t\hat{\sigma}_e$ and $\tilde{x} + c_t\hat{\sigma}_e$, where c_t is the appropriate constant as given in tables of the t-distribution, using t as the 'degrees of freedom'.

The importance of attempting to give a confidence interval for a forecast cannot be exaggerated. To state, in the above example, that the forecast is 245 items, without any indication of its possible accuracy, is of little use. In the example, the 95 per cent confidence limits were (224.2, 265.8) when σ was known. Suppose, however, they had been (0, 490); then our attitude to the forecast of 245 would have been entirely different. Table 4.3 gives a summary of the main results of this section. Notice that in $\tilde{x}_{t,k}$ all observations carry equal weight. The earliest observation, x_1, is as important as the latest, x_t. This is a consequence of the fact that we have treated our basic model as being globally valid. We have used it as a global constant mean model.

4.2 The local constant mean model

The basic structure for the local constant mean model is exactly that given in the last section. However, it is now believed that, though this model is a good one for short periods of time, it is unlikely that the value of μ for one locality in time is exactly the same as that for another. Thus μ is regarded as a slowly varying quantity. Later we shall

Table 4.4 Calculation of moving averages

Month	1	2	3	4	5	6	7	8	9	10	11	12
Sales	7	5	6	4	5	4	5	3	4	7	6	5
First set	7	5	6	4	5							
Average			5.4									
Second set		5	6	4	5	4						
Average				4.8								
						etc.						
Eighth set								3	4	7	6	5
Average										5.0		

Moving averages	5.4	4.8	4.8	4.1	4.2	4.6	5.0	5.0

consider ways of modelling this variation in μ, but for now we shall assume that it might happen in a slow but unspecified way. In the global constant mean model the averaging operation over all the data had the effect of cutting down the random variation and leaving an estimate of μ. If an estimate of μ is required in just one locality of the data, then it is reasonable to average the observations just from that locality and ignore the rest. In Table 4.4 some illustrative figures indicate how an average can be obtained from any five-month period in the total set of data. Thus if we wish to estimate μ for the most recent period of, say, 5 months, we simply average the last five observations. Hence the estimate is $(3 + 4 + 7 + 6 + 5)/5 = 5.0$. This would, at this simple level, be the estimate of μ that could be used to forecast x_{14}. If the mean, μ, was wandering to any extent in the time covered by the five observations, this forecast would obviously be a biased forecast. In this situation all we could say would be that the value of the average, 5.0, is an estimate of μ at the middle of the five observations, i.e. at week 10. Similarly, the averages of each successive group of five can be regarded as estimates of the mean at the centre of their ranges. Thus the last line in Table 4.4 gives the estimates of the means for each month found by moving the sequence of five numbers successively to the right by one. The averages so obtained are thus termed 'moving averages'. It will be seen that there are less moving averages than observations, and the latest value of the mean estimated will always be that corresponding to some time before the latest observation. Where an average of $2n + 1$ observations is used, there will be n observations after the time at which the latest moving average is centred. If there are an even number of observations, the centre of the range of the moving average will fall between two observations. To obtain values that estimate the means at observations times, these moving averages are themselves averaged in pairs.

If a moving average of s observations is used, the forecast of x_{t+k} would be

$$\tilde{x}_t = \frac{x_{t-s+1} + \ldots + x_t}{s}$$

When new observations are obtained, a recurrence relation can be used since

$$\tilde{x}_{t+1} = (x_{t+1} + s\tilde{x}_t - x_{t-s+1})/s$$

Both these forms involve storage of the last s observations. If a moving average is to be seriously considered as a means of forecasting, there must be good evidence that the mean is only wandering slightly. A somewhat different use for moving averages in forecasting will be considered in a later chapter.

As already pointed out, a moving average is essentially an estimate of the mean at the mid-time of the time interval covered. Its main practical use is thus to smooth the data rather than forecast the future. As pointed out by Davis (1941), when the interest is in forecasting it would be more reasonable to use a moving average which gives more weight to recent data than past data. For example, instead of a simple moving average

$$\frac{1}{4}(x_{t-3} + x_{t-2} + x_{t-1} + x_t)$$

we might use

$$\frac{1}{10}(x_{t-3} + 2x_{t-2} + 3x_{t-1} + 4x_t)$$

This would give four times as much attention to x_t as to x_{t-3}, instead of the equal attention given by the ordinary moving average. This would intuitively make it more suitable for forecasting purposes. The divisor, 10, is introduced as being the sum of the weights, $1 + 2 + 3 + 4$. The reason for this is that if, for example,

$$x_{t-3} = x_{t-2} = x_{t-1} = x_t = 7$$

then this 'weighted' moving average will also have the value 7. Though this weighted moving average estimates μ locally, in the sense that it takes most notice of current information, it suffers from two disadvantages. Firstly, it only uses the four latest observations and ignores the rest. Secondly, it does not possess a very simple recurrence form. A set of weights for a weighted moving average which overcomes both these disadvantages is the 'exponential' set of weights given by

Observation	x_1	x_2	x_{t-2}	x_{t-1}	x_t
Weight	a^{t-1}	a^{t-2}	a^2	a^1	a^0

where a is a constant with value $0 < a \leqslant 1$. The general formula for the exponentially weighted moving average is

$$\tilde{x}_t = \frac{\displaystyle\sum_{r=0}^{t-1} a^r x_{t-r}}{\displaystyle\sum_{r=0}^{t-1} a^r}$$

Table 4.5(a) shows the calculation of this weighted mean directly as in the formula. This looks very complicated until one realizes that each week's new values can be calculated from the previous week's values. Thus the weighted sum of the observations, the numerator, for week 3 is

$$0.8^2 \times 7 + 0.8 \times 5 + 1.0 \times 6$$

which is

$$0.8 \, (0.8 \times 7 + 5) + 6$$

which is

$$0.8 \text{ (weighted sum for week 2)} + \text{latest observation}$$

similarly, the sum of weights, the denominator, for week 3 is

$$0.8^2 + 0.8 + 1.0$$

which is

$$0.8 \, (0.8 + 1.0) + 1.0$$

which is

$$0.8 \text{ (sum of weights for week 2)} + 1.0$$

We can thus replace the above formula by a simple set of calculations as follows. Denote the numerator of $\tilde{x}_t$ by S_t and the denominator by W_t, i.e.

$$S_t = \sum_{r=0}^{t-1} a^r x_{t-r}$$

and

$$W_t = \sum_{r=0}^{t-1} a^r$$

Table 4.5(a) Calculations for exponentially weighted mean, explicit form $a = 0.8$

	Week number	1	2	3	4
Week	Sales	7	5	6	4

1	Weight	1.0
	Weighted sum	$1.0 \times 7 = 7$
	Sum of weights	1.0
	$\tilde{x}_1$	$7/1.0 = 7$

2	Weights	0.8 1.0
	Weighted sum	$0.8 \times 7 + 1.0 \times 5 = 10.6$
	Sum of weights	$0.8 + 1.0 = 1.8$
	$\tilde{x}_2$	$10.6/1.8 = 5.9$

3	Weights	0.8^2 0.8 1.0
	Weighted sum	$0.8^2 \times 7 + 0.8 \times 5 + 1.0 \times 6 = 14.48$
	Sum of weights	$0.8^2 + 0.8 + 1.0 = 2.44$
	$\tilde{x}_3$	$= 14.48/2.44 = 5.93$

Table 4.5(b) Tabular layout for exponentially weighted mean, $a = 0.8$

(1)	(2)	(3)	(4)	(5)	(6)
Time t	Observation x_t	$0.8 \times$ previous s	(2) + (3) S_t	Divisor $W_t = \Sigma a^r$	(4)/(5) $\tilde{x}_t$
1	7	0.0	7.0	1.0	7.0
2	5	5.6	10.6	1.8	5.89
3	6	8.48	14.48	2.44	5.93
4	4	11.58	15.58	2.95	5.28
5	5	12.47	17.47	3.36	5.20
6	4	13.97	17.97	3.69	4.87
7	5	14.38	19.38	3.95	4.91
8	3	15.50	18.50	4.16	4.45
9	4	14.80	18.80	4.33	4.34

Then

$$S_t = aS_{t-1} + x_t$$
$$W_t = aW_{t-1} + 1$$
$$\tilde{x}_t = S_t/W_t$$
$$S_0 = 0, \; W_0 = 0$$

Thus both numerator and denominator may be evaluated using very simple recurrence formulae. This is illustrated in Table 4.5(b). In moving from one stage to the next of this calculation, only the last values of S_t and W_t need to be recorded.

One way of looking at the introduction of the weights a^r is that we now look at our data through a fog. When a is small the fog is thick, so that we can only see with any clarity the data that are close to us in time. Putting $a = 1$ corresponds to a clear day in which we can see all the data clearly. Technically, the association of weights with variables to decrease their importance is called discounting and we will refer to a as the discounting factor.

Let us now consider two other approximate ways of expressing our forecasting formula. If $a < 1$ and if t is so large that a^t is negligible, then the value of W_t can be shown to tend towards a constant limit of $(1-a)^{-1}$ and thus only S_t needs to be stored. In this case the formula becomes

$$\tilde{x}_t = (1-a) \sum_{r=0}^{t-1} a^r x_{t-r}$$

This can be rewritten in two ways that are very convenient for calculation purposes:

$$\tilde{x}_t = (1-a)(x_t + ax_{t-1} + a^2 x_{t-2} + \ldots)$$
$$= (1-a)x_t + a(1-a)(x_{t-1} + ax_{t-2} + \ldots)$$

Thus

$$\tilde{x}_t = (1-a)x_t + a\tilde{x}_{t-1}.$$

This recurrence relation, as that for the ordinary average in section 4.1, relates the new forecast to the old forecast and the new observation. The values a and $1-a$ give the proportion of weight attached to the old forecast and new observation, respectively. In words, we may thus write

New forecast $= (1-a) \times$ new observation $+ a \times$ old forecast

This recurrence relation can be treated as a forecasting formula in its own right, instead of as an approximation, and is referred as 'exponential smoothing' (e.g. Brown, 1959; Winters, 1960). The fact that the weights are constant enable a simple nomogram to be used for updating. This is shown in Figure 4.2.

We may also rewrite the recurrence equation in terms of forecast errors. Writing

$$e_t = x_t - \tilde{x}_{t-1}$$

the last equation becomes

$$\tilde{x}_t = (1-a)(e_t + \tilde{x}_{t-1}) + a\tilde{x}_{t-1}$$

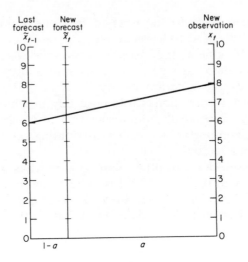

Figure 4.2   A nomogram for exponential smoothing

Hence

$$\tilde{x}_t = \tilde{x}_{t-1} + (1-a)e_t$$

Thus the new forecast can be obtained from the old by adding a fraction $(1-a)$ of the last error made in our forecasting. The only difference between this and the similar expression for the global mean model is that the fraction is a constant independent of t rather than a fraction which gets smaller as t increases. Intuitively, this means that the forecast is always 'constantly' alert for changes in the situation which would reveal themselves through the forecast errors. This is exactly the feature that is required if we are to forecast a constant mean model which is valid locally rather than globally. The method has been introduced to deal with a model approximating to the global constant mean model

$$x_t = \mu + \epsilon_t$$

over small localities of time, but for which the mean μ wanders in some unspecified fashion. Many theoretical models have been devised that give, in effect, wandering means. It has been shown that for many of these the exponentially weighted moving average gives a good forecast. Clearly, if we can be more precise about how the mean wanders, we should be able to improve on this forecast. We will discuss suitable methods in later chapters. Experience shows, however, that this method, often called exponential smoothing, provides a very robust and widely applicable technique.

As a means of studying the method, let us compare it with the method discussed in the previous section, assuming the 'true' model to be the global constant mean model. Consider the expectation and variance of $\tilde{x}_t = S_t/W_t$. For a geometric series, which our so-called 'exponential' weights actually form,

$$W_t = 1 + a + a^2 + \ldots + a^{t-1} = (1 - a^t)/(1 - a)$$

and from the model

$$S_t = \mu \sum_{r=0}^{t-1} a^r + \sum_{r=0}^{t-1} a^r \epsilon_{t-r}$$

Hence

$$E(S_t) = \mu W_t + 0$$

and

$$E(\tilde{x}_t) = \mu$$

Thus exponential smoothing leads to an unbiased forecast if the mean is genuinely constant. Further,

$$\text{Var}(S_t) = \sum_{r=0}^{t-1} a^{2r} \sigma^2 = \frac{1 - a^{2t}}{1 - a^2} \sigma^2$$

But

$$\text{Var}(\tilde{x}_t) = \text{Var}(S_t)/W_t^2$$

and so after simplifying

$$\text{Var}(\tilde{x}_t) = \frac{1 - a}{1 + a} \cdot \frac{1 + a^t}{1 - a^t} \sigma^2$$

For t large this becomes

$$\text{Var}(\tilde{x}_t) = \frac{1 - a}{1 + a} \sigma^2 \quad (0 < a < 1)$$

This variance is fundamentally different in form to that obtained for the forecast in the previous section, σ^2/t, which becomes smaller as t is increased and in fact approaches zero. For exponential smoothing the variance of $\tilde{x}_t$ approaches the minimum value given above and however much data we have we shall never get a smaller variance than this. The reason is intuitively clear since as new data are added less and less notice is taken of old data, so the effective amount of data available remains constant when t is large. It will also be noticed that as a, the discounting factor, gets closer to one, for any fixed t, $\text{Var}(\tilde{x})$ gets smaller and a plot of $\tilde{x}_t$ will look smoother. This occurs because, as we

56

Table 4.6 Some values of $\text{Var}(\tilde{x}_t)/\sigma^2$ for constant mean models

| | $t = 10$ | | $t = \infty$ | |
	Global model	Local model	Global model	Local model
$\tilde{x}_t$ = average of x s	0.10	1.27	0	∞
$\tilde{x}_t$ = exponential smoothing	0.14	0.19	0.11	0.12

saw in the previous section, the best forecasting formula for this model is the ordinary average which corresponds to the exact form of exponential smoothing with $a = 1$. By comparing values of $\text{Var}(\tilde{x})$ for various values of a, including $a = 1$, we can see how much we lose by using our exponential weights when we are in fact dealing with a global constant mean model. Conversely, if we consider some model for a wandering mean process and repeat the comparison, we can see the advantage of introducing the weights a^r. Table 4.6 shows such a comparison for a particular wandering mean process. Notice from the table how little is lost, by way of having a larger variance, when $a = 0.9$, say, is used instead of $a = 1.0$ in the global constant mean model, and conversely how much bigger the variance is if $a = 1.0$ is used when some $a < 1.0$ should have been used. It would appear from this example that unless we can be absolutely certain that the constant mean model has applied in the past, *and will continue to apply in the future*, then it will be safer to use the weighted mean rather than the ordinary mean.

In the previous section we discussed the construction of confidence limits for the future observation. It might appear reasonable to use the above expressions for variance to construct such limits, and indeed we could. The limits, however, would refer to the permanent model; the limits for a local model would, in general, be different. The natural approach here is to estimate σ_e directly from the observed forecast errors using the estimator $\hat{\sigma}_e$ given in the last section. The prediction

Table 4.7 Local constant mean model — exponential smoothing

Model	$x_t = \mu + \epsilon_t$, $t = 0, 1, \ldots$, ϵ_t independent, $E(\epsilon_t) = 0$, $\text{Var}(\epsilon_t) = \sigma^2$ Model regarded as a local approximation
Explicit form	$\tilde{x}_t = \sum_{r=0}^{t-1} a^r x_{t-r} / \sum_{r=0}^{t-1} a^r$, independent of lead time k
Recurrence form	$\tilde{x}_t = (1-a)x_t + a\tilde{x}_{t-1}$, t large, $a < 1$
Error correction form	$\tilde{x}_t = \tilde{x}_{t-1} + (1-a)e_t$, t large, $a < 1$

intervals can then be derived using the same approach as that discussed in that section.

The forecasting formula of exponential smoothing involves the 'discounting factor' a, which is a 'forecasting parameter' as distinct from a parameter of the model. The value of a is normally chosen so as to minimize the mean square error of forecasts over a trial period. This is discussed in more detail in Chapter 16. Table 4.7 gives a summary of the main formulae for this section. It should be noted that much of the literature on exponential smoothing uses $\alpha = 1 - a$ as the basic parameter. This is often referred to as the 'smoothing constant'.

References

Brown, R. G. (1959). *Statistical Forecasting for Inventory Control.* McGraw-Hill, New York.

Davis, H. T. (1941). *The Analysis of Time Series.* Cowles Commission, Yale.

Fisher, R. A. and Yates, F. (1938). *Statistical Tables.* Oliver and Boyd, Edinburgh.

Hogg, R. V. and Craig, A. T. (1970). *Introduction to Mathematical Statistics.* Macmillan, New York.

Mood, A. M. and Graybill, F. A. (1963). *Introduction to the Theory of Statistics.* McGraw-Hill, New York.

Winters, P. R. (1960). Forecasting sales by exponentially weighted moving averages. *Man. Sci.,* 6, 324—342.

Chapter 5

Linear trend models

5.1 The global linear trend model

We turn now to the problem of forecasting future values of a structure which shows a linear trend with added random variation. Thus the model is

$$x_t = \alpha + \beta t + \epsilon_t$$

where α is the expectation of x_0, β is the slope and ϵ is a sequence of independent random variables with $E(\epsilon_i) = 0$, $\text{Var}(\epsilon) = \sigma^2$ for all i, and whose distribution will, for the moment, remain unspecified. The data in Table 5.1, charted in Figure 5.1, are typical of data for which such a model might be reasonable. If this data did arise from the model, then the model describes the future value x_{12} as

$$x_{12} = \alpha + \beta.12 + \epsilon_{12}$$

As we have assumed the ϵs to be independent of each other, we have no way of forecasting ϵ_{12}. All we can do is to set ϵ_{12} at its expected value of zero. Thus if we knew α and β, our forecast would be $\alpha + \beta.12$. Unfortunately, we do not know α or β so we must use our data to find values for them. In this section we shall investigate several ways of estimating the values of α and β from a set of data and hence of extrapolating the line to obtain forecasts of a future x_t.

5.1.1 Fitting by eye

When a set of data is replaced by a corresponding curve, we say that we have fitted the curve to the data and the curve is called the fitted curve. In fitting a straight line to data such as that of Figure 5.1 we can simply lay a transparent ruler over the plotted data and adjust it until the data is scattered evenly and approximately equally about the edge of the ruler. Figure 5.2 shows such a line for the data of Table 5.1. The ease with which we can do this depends on how close the data looks to

Table 5.1 Data from a global linear trend model

Time	1	2	3	4	5	6	7	8	9	10	11
x	31	34	37	36	42	39	42	45	47	47	52

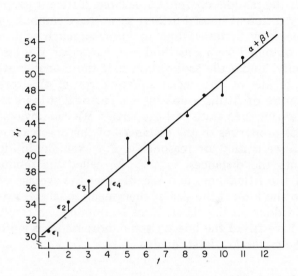

Figure 5.1   Plot of the data of Table 5.1

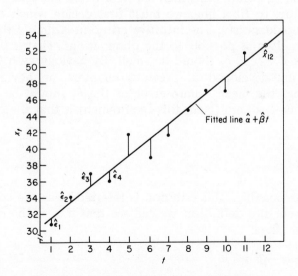

Figure 5.2   Fitting a line and extrapolating

a straight line. This is obviously a very easy way of obtaining a line. It has the advantage that if we know that certain of the points are suspect, in the sense of not being known as accurately as the others, or that they are produced by known and exeptional events, then we can take less notice of these points in assessing the line. However, this subjective element is also the weakness of the method for, in data where the variability of the random component is large, different people may well produce very different lines. Further, as we have emphasized before, it is always necessary in forecasting to know as much as possible about the forecast obtained from a method — in particular, it is essential to be able to specify, under the assumption that the model continues to be valid, an estimate of the mean square error of the forecasts. The subjective nature of fitting a line by eye means that there is no built-in means of assessing such statistical properties. We can obviously examine the statistical properties of the distances of the observations from the fitted line to see if they are reasonable. For example, in Figure 5.2 we could measure the distances, $\hat{\epsilon}_1, \ldots, \hat{\epsilon}_t$, called the residuals, of the points from the fitted line, and see whether the average of these was close to zero and look at the size of their mean square. However, by the time we have done all the calculations to obtain such results we might just as well have fitted the line by some more rigorous method. Such a method is *the method of least squares*.

5.1.2 The method of least squares

Refer again to Figure 5.2 which shows a fitted line and the data. To decide on how to fit a line, we must first decide which criteria the fitted line should satisfy. The intuitive criterion is simply that the line should be as close as possible to the observations. That is to say, that the residuals $\hat{\epsilon}_1, \ldots, \hat{\epsilon}_t$ should be small. By analogy with the criteria for forecasting developed in previous chapters, an obvious criteria would be that the mean square error of the $\hat{\epsilon}_i$ should be as small as possible. Since t is here fixed, this requirement is that the quantity S, where

$$S = \sum_{i=1}^{t} \hat{\epsilon}_i^2$$

is as small as possible. This criterion is referred to as the least squares criterion. From the definition we can see that if x_i is the observation made at time i then

$$\hat{\epsilon}_i = x_i - \hat{\alpha} - \hat{\beta}i$$

where $\hat{\alpha}$ and $\hat{\beta}$ are fitted values, the estimators, of α and β respectively.

Thus

$$S = \sum_{i=1}^{t} (x_i - \hat{\alpha} - \hat{\beta}i)^2$$

The least squares line is the line for which $\hat{\alpha}$ and $\hat{\beta}$ makes S a minimum. A direct application of calculus (setting $\partial S/\partial\hat{\alpha} = 0$ and $\partial S/\partial\hat{\beta} = 0$) gives two simultaneous equations for $\hat{\alpha}$ and $\hat{\beta}$. These are

$$\sum_{i=1}^{t} x_i = \hat{\alpha}t + \hat{\beta} \sum_{i=1}^{t} i$$

$$\sum_{i=1}^{t} ix_i = \hat{\alpha} \sum_{i=1}^{t} i + \hat{\beta} \sum_{i=1}^{t} i^2$$

These equations are referred to as the 'normal' equations. If the times at which the observations are taken are not simply $1, 2, \ldots, t$, then the summations refer to all the times at which the actual observations were made. Notice the relation of these equations to the equation of the fitted line:

(a) The fitted line is $x_i = \hat{\alpha} + \hat{\beta}i$;
(b) summing both sides over all observations gives the first normal equation;
(c) multiplying both sides by the time, i, and then summing gives the second normal equation.

It sometimes helps in solving these equations to use artificial time origins. Table 5.2 gives an example in which $i = 6$ is taken as a new time origin from which time i' is measured. As $\Sigma i' = 0$, the first normal equation becomes $\hat{\alpha} = \bar{x}$, the average value of x, and the second gives $\hat{\beta} = \Sigma i'x/\Sigma i'^2$. Note that here α is the expected value of x at time $i' = 0$. Croxton, Cowden and Klein (1968) discuss various ways of simplifying the solution of normal equations. Notice that if we know β to be zero, for the constant mean model, the result $\hat{\alpha} = \bar{x}$ shows that the estimate of μ used in the previous chapter was in fact the least squares estimate.

If we keep to an origin for which the sum of the observation times is zero, i.e. $i' = i -$ average time i, and we substitute for x_i the value $\alpha + \beta i' + \epsilon_i'$, from the model, then we can show that:

(a) $E(\hat{\alpha}) = \alpha, E(\hat{\beta}) = \beta$;
(b) $\text{Var}(\hat{\alpha}) = \sigma^2/t$, $\text{Var}(\hat{\beta}) = \sigma^2/\Sigma i'^2$, t still being used to denote the number of observations;
(c) $\hat{\alpha}$ and $\hat{\beta}$ are independent.

The forecast of any future value of x, say x_T, is simply obtained by substituting the time T in the fitted line; thus

$$\hat{x}_T = \hat{\alpha} + \hat{\beta}T$$

Table 5.2 Calculation of least squares line

i	x	i'	i'^2	$i'x$	$\hat{\alpha} + \hat{\beta}t'$	$\hat{\epsilon}$	$\hat{\epsilon}^2$
1	31	−5	25	−155	31.68	−0.68	0.462
2	34	−4	16	−134	33.56	0.44	0.194
3	37	−3	9	−111	35.44	1.56	2.434
4	36	−2	4	−72	37.33	−1.33	1.769
5	42	−1	1	−42	39.21	2.79	7.784
6	39	0	0	−514	41.09	−2.09	4.368
7	42	1	1	42	42.97	−0.97	0.941
8	45	2	4	90	44.85	0.15	0.023
9	47	3	9	141	46.74	0.26	0.068
10	47	4	16	188	48.62	−1.62	2.624
11	52	5	25	260	50.50	1.50	2.250
				721			
	452	0	110	207			22.916

$x_{i'} = \alpha + \beta i'$ $\hat{\alpha} = \bar{x} = 41.09$

$\Sigma i' = 0$ $\hat{\beta} = \Sigma x i'/\Sigma i'^2 = 1.882$

∴ Fitted line = 41.09 + 1.882 $(i - 6)$ = 29.80 + 1.88 i

Estimated variance $\hat{\sigma}^2 = \Sigma \hat{\epsilon}^2/9 = 1.59^2$

Forecast of x_{12} = 41.09 + 1.882 × 6 = 52.38

The forecast error is

$$e_T = x_T - \hat{x}_T$$
$$= (\alpha + \beta T + \epsilon_T) - (\hat{\alpha} + \hat{\beta}T)$$
$$= \alpha - \hat{\alpha} + (\beta - \hat{\beta})T + \epsilon_T$$

Hence $E(e_T) = 0$, from (a) above, so the forecast is unbiased. It follows that the mean square error,

$$\text{M.S.E.} = E(e_T^2)$$
$$= \text{Var}(e_T)$$
$$= \text{Var}(\hat{\alpha}) + T^2 \, \text{Var}(\hat{\beta}) + \sigma^2$$

Thus

$$\text{M.S.E.} = \sigma^2 \left(\frac{1}{t} + \frac{T^2}{\Sigma i'^2} + 1 \right)$$

It will be seen that as T, the time from the average time of the observations to the time being forecast, increases so does the mean

Table 5.3 Forecasts and limits for data of Table 5.1

(a) Known $\sigma^2 = 2.5$

Time	$\hat{x}_T$	σ_{e_T}	95% prediction interval		
12	52.380	1.883	48.689	to	56.071
13	54.266	1.960	50.424	to	58.108
14	56.148	2.045	52.140	to	60.156
15	58.030	2.137	53.841	to	62.218
16	59.912	2.237	55.527	to	64.297

(b) Estimated $\hat{\sigma}^2 = 2.546$ using 9 degrees of freedom

Time	$\hat{x}_T$	$\hat{\sigma}_{e_T}$	95% prediction interval		
12	52.380	1.900	48.082	to	56.678
13	54.266	1.978	49.792	to	58.740
14	56.148	2.064	51.479	to	60.817
15	58.030	2.157	53.151	to	62.909
16	59.912	2.257	54.807	to	65.017

Note $\hat{\sigma}^2 = \Sigma \hat{\epsilon}_i^2 /(t-2)$, with $t-2$ degrees of freedom.
The interval is

$$(\hat{x}_T - c_{t-2}\hat{\sigma}_{e_T}, \hat{x}_T + c_{t-2}\hat{\sigma}_{e_T})$$

and

$$\hat{\sigma}_{e_T} = \hat{\sigma}^2 \left(\frac{1}{t} + \frac{T^2}{\Sigma i'^2} + 1 \right)$$

square error. Thus if we use this as σ_e^2 we can derive prediction intervals for $\hat{x}_T$ as we did in Chapter 4. However, this interval gets wider as T increases. This is illustrated in Table 5.3. In deriving our forecast we have chosen the least squares line and extrapolated. It is possible to show that this approach leads to the forecast with the smallest mean square error. This is not true for all models, but it is true with models that are linear in their parameters.

5.2 The local linear trend model

In the same way as the permanency of a mean may be open to doubt so, too, may the permanency of a trend line. So we consider in this section the forecasting of data which is not adequately described by a global trend model but for which trend lines provide reasonable models for each locality of time. Table 5.4 gives a set of sales data for which a local trend model seems appropriate. Figure 5.3 shows a plot of this data. As in section 4.2 it might be advisable to investigate more

Table 5.4 Data showing local linear trends

t	x_t	t	x_t	t	x_t	t	x_t	t	x_t
1	42.8	11	40.8	21	38.8	31	38.1	41	35.8
2	42.6	12	40.5	22	38.8	32	37.8		
3	42.5	13	40.2	23	38.8	33	38.1		
4	42.4	14	39.9	24	38.6	34	37.7		
5	42.1	15	39.6	25	38.4	35	37.5		
6	41.8	16	39.5	26	38.4	36	37.2		
7	41.5	17	39.5	27	38.4	37	37.0		
8	41.3	18	39.2	28	38.5	38	36.8		
9	41.1	19	39.1	29	38.4	39	36.5		
10	40.9	20	38.9	30	38.2	40	36.2		

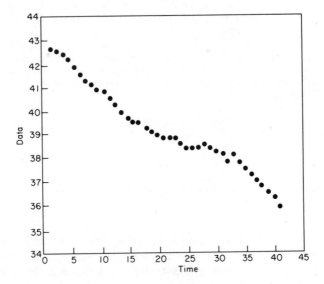

Figure 5.3   Plot of the data of Table 5.4

sophisticated models, but often a local linear trend will provide a simple and robust model as the basis for forecasting. The terms in the more sophisticated model might well describe the past more accurately than our local trend model, but we may have doubts about their continuing stability in the future. It will often seem in practice that the linear trend is the only stable aspect of the situation that one can reasonably expect to continue into the future, at least for a short time. This distinction in attitude to past and future must always be borne in mind by the forecaster. What he regards as a good model will often be different from that of the person whose only aim is to obtain a good model for a given set of data from the past. We shall consider in this section three approaches to forecasting with this model.

5.2.1 Holt's method

In discussing exponential smoothing the estimate '$\tilde{x}_t$' of the underlying mean was expressed as a recurrence relation. There we used the notation $\tilde{x}_t$ for the estimate of μ, which was also the forecast of the future observation x_{t+h}. To emphasize that the equation did refer to estimates, $\hat{\mu}_t$, of the current mean, we shall rewrite the recurrence relation as

$$\hat{\mu}_t = (1 - a)x_t + a\hat{\mu}_{t-1}$$

This is essentially a weighted mean, with weights $(1-a)$ and a of the latest data x_t, which estimates the current mean and the old estimate $\hat{\mu}_{t-1}$. In Holt's method (see Coutie and coauthors, 1966; Holt, 1957) we seek to produce corresponding recurrence relations for the estimates of the two parameters in the trend model. However, rather than refer to the parameter α, which was the mean of the model at time zero, we rewrite the model so that α is replaced by μ_t, the mean of the process at the present time t. Thus at time $t-r$ the model is

$$x_{t-r} = \mu_t - \beta r + \epsilon_{t-r}$$

where r is time-measured into the past. By its definition and the fact that the slope is β, it is clear from Figure 5.4(a) that adjacent values of μ are related by

$$\mu_t = \mu_{t-1} + \beta$$

Hence an estimate of μ_t can be found from the two previous estimates of μ_{t-1} and β by simply adding $\hat{\beta}_{t-1}$ to $\hat{\mu}_{t-1}$, see Figure 5.4(b). This gives the estimate of the mean at time t based on past data. The latest observation x_t also provides an estimate of μ_t. Hence if we use the ideas in the exponential smoothing recurrence equation and make the new estimate $\hat{\mu}_t$ a weighted mean of the latest data and the estimate based on using the previous estimates, we arrive at

$$\hat{\mu}_t = (1 - a)x_t + a(\hat{\mu}_{t-1} + \hat{\beta}_{t-1})$$

To make use of this equation, we need a way of calculating the sequence of estimates $\hat{\beta}_{t-1}$. This may be done by applying the idea used in getting the above recurrence relation to finding a recurrence relation for the β_t values. The new value β_t will be a weighted mean of the old estimate $\hat{\beta}_{t-1}$ and an estimate based on the most recent data. To estimate the slope using the most recent data, we might use $x_t - x_{t-1}$, since from the model this can be written as

$$x_t - x_{t-1} = (\mu_t + \epsilon_t) - (\mu_{t+1} - \beta + \epsilon_{t-1})$$
$$= \beta + \epsilon_t - \epsilon_{t-1}$$

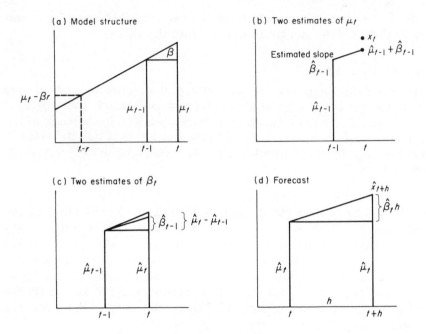

Figure 5.4   The basis of Holt's method

Thus $x_t - x_{t-1}$ has expectation β and thus provides an estimate of β. Unfortunately, because it involves both ϵ_t and ϵ_{t-1}, it has a large variance and so the estimates of β would not be very stable. Figure 5.4(c) suggests a better way of estimating. This is to use the relation

$$\beta = \mu_t - \mu_{t-1}$$

By replacing the unknown parameters μ_t and μ_{t-1} by our estimates of them, we see that $\hat{\mu}_t - \hat{\mu}_{t-1}$ gives an estimate of β and includes implicitly the influence of the latest observation x_t. It is, however, not as strongly influenced by x_t as the previously suggested estimate. The method of updating the estimate of slope is thus based on the recurrence relation

$$\beta_t = (1-b)(\hat{\mu}_t - \hat{\mu}_{t-1}) + b\hat{\beta}_{t-1}$$

Note that we do not have to use the same weighting constant as for $\hat{\mu}_t$, so a new constant b is introduced.

The practical application of these formula requires the estimates of β and μ to be updated alternately using the recurrence formulae above, each updated estimate being used in the next calculation of the other parameter. The calculation is illustrated in Table 5.5. We now have a way of finding estimates of the slope and current mean at the time of

Table 5.5 Example of Holt's method

(a) Formulae:

$$\hat{\mu}_t = (1-a)x_t + a(\hat{\mu}_{t-1} + \hat{\beta}_{t-1})$$
$$\hat{\beta}_t = (1-b)(\hat{\mu}_t - \hat{\mu}_{t-1}) + b\hat{\beta}_{t-1}$$
$$\hat{x}_{t+2} = \hat{\mu}_t + 2\hat{\beta}_t$$
$$a = 0.8 \quad b = 0.9$$

Initial values assumed $\hat{\mu} = 42.9, \hat{\beta} = -0.15$

(b) Data from Table 5.4

0.2	0.8		0.1	0.9		
x_t	$\hat{\mu}_{t-1} + \hat{\beta}_{t-1}$	$\hat{\mu}_t$	$\hat{\mu}_t - \hat{\mu}_{t-1}$	$\hat{\beta}_{t-1}$	$\hat{\beta}_t$	$\hat{x}_{t+2}$
—	—	42.900	—	—	−0.15	—
42.8	42.75	42.760	−0.140	−0.150	−0.149	42.46
42.6	42.611	42.609	−0.151	−0.149	−0.149	42.311
42.5	42.459	42.468	−0.141	−0.149	−0.148	42.172
42.4	42.320	42.336	−0.132	−0.148	−0.146	42.044
42.1	42.190	42.172	−0.164	−0.146	−0.148	41.876
41.8	42.024	41.979	−0.193	−0.148	−0.152	41.675
41.5	41.826	41.761	−0.218	−0.152	−0.159	41.443

the most recent observation. The estimated equation of the line at time t is

$$x_{t-r} = \hat{\mu}_t - \hat{\beta}_t r \quad (r = 0, 1, \ldots, t-1)$$

A forecast of x_{t+h} is thus provided by

$$\hat{x}_{t,h} = \hat{\mu}_t + \hat{\beta}_t h$$

obtained by extrapolating along the estimated line as in Figure 5.4(d). As in the constant mean model, the independence condition on the ϵ values means that we can only forecast the deterministic part of the situation and not the future random variable ϵ_{t+h}.

To start forecasting using recurrence relations, we need good starting values $\hat{\mu}_0$ and $\hat{\beta}_0$, see Figure 5.5. These values are best based on an initial set of data. From this set of data an initial line can be obtained by any of the methods of sections 5.1. $\hat{\beta}_0$ is then taken as the slope of this line and $\hat{\mu}_0$ as its value at the time one unit before the initial observation is obtained.

If there is an initial set of data available, it is wise to use it to obtain some idea as to the best values for the forecasting parameters a and b. For example, if we take the mean square error as the criterion, then we

68

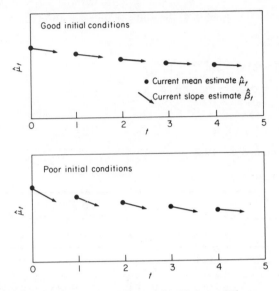

Figure 5.5   The effects of initial conditions

can evaluate the mean square error for a range of different values of a and b. Since all methods produce poor forecasts when there is very little data, the forecasts obtained over the first few observations should be ignored in calculating the mean square error. A discussion of methods of choosing parameter values is given in section 15.3.

In Chapter 4 we noted that the recurrence form of forecasting formula could also be written in terms of the forecast errors. This can also be done here. The one-step-ahead forecast error is

$$e_t = x_t - (\hat{\mu}_{t-1} + \hat{\beta}_{t-1})$$

Substituting the value of x_t given by this definition in the recurrence relation for $\hat{\mu}_t$ gives

$$\hat{\mu}_t = (1-a)\{e_t + (\hat{\mu}_{t-1} + \hat{\beta}_{t-1})\} + a(\hat{\mu}_{t-1} + \hat{\beta}_{t-1})$$

Hence

$$\hat{\mu}_t = \hat{\mu}_{t-1} + \hat{\beta}_{t-1} + (1-a)e_t$$

which is the obvious extension to a trend line of the error correction form of Chapter 4. If we substitute the expression of $\hat{\mu}_t - \hat{\mu}_{t-1}$ from this into the recurrence relation for $\hat{\beta}_t$, we get

$$\hat{\beta}_t = (1-b)\{\hat{\beta}_{t-1} + (1-a)e_t\} + b\hat{\beta}_{t-1}$$

Hence

$$\hat{\beta}_t = \hat{\beta}_{t-1} + ce_t$$

where $c = (1 - b)(1 - a)$. Thus we can update both the mean and slope estimates by adding simple multiples of e_t.

5.2.2 Discounted least squares

In the discussion on the least squares method on section 5.1.2 it was stated that it was reasonable to choose the parameters in the model to minimize the sum of the squares of the distances of the observations from the line that we propose to fit to the data. This sum of squares we denoted by

$$S = \sum_{t=1}^{t} \hat{e}_i^2$$

where the $\hat{e}$ values were called the residuals. In doing this we attach as much importance to residual $\hat{e}_1$, for our very first observation, as to the residual $\hat{e}_t$, for our latest observation. When we are dealing with data believed to come from a local model and our aim is to get a good fit to the recent data for forecasting purposes, then this equality of importance may not be very sensible. It may be more reasonable to introduce a weighting, or discounting, into the sum of residuals squared in exactly the same way that we introduced it into the ordinary mean in justifying use of the weighted mean of exact exponential smoothing. Various types of weighting or discounting could be introduced but, as in the constant mean model, it is found that the use of exponential weights lead to particularly simple calculations. We thus replace the sum of residuals squared by a weighted sum in which $\hat{e}_{t-r}^2$ has a weight of a^r $(0 < a \leqslant 1)$ associated with it. The most recent observation gives a residual squared with weight one. The far-past residuals, having small values of a^r (for $a < 1$), contribute little to the total. As the magnitude of a controls the extent to which the past is discounted, it is often called the 'discounting factor'. Thus our least squares criterion is replaced by

$$S = \sum_{r=0}^{t-1} a^r \hat{e}_{t-r}^2$$

which is the *discounted least squares* criterion (see Gilchrist, 1967). We can now substitute the value of $\hat{e}_{t-r}$ given by our line, again taking the present as origin for r, i.e.

$$\hat{e}_{t-r} = x_{t-r} - \hat{\mu}_t + \hat{\beta}_t r$$

We now choose the values of $\hat{\mu}_t$ and $\hat{\beta}_t$ to minimize S. This is done in detail in Appendix D. The solution can be put in the form of normal equations, as in section 5.1, the new normal equations being

$$\hat{\mu}_t \sum_{r=0}^{t-1} a^r - \hat{\beta}_t \sum_{r=0}^{t-1} ra^r = \sum_{r=0}^{t-1} a^r x_{t-r}$$

$$\hat{\mu}_t \sum_{r=0}^{t-1} ra^r - \hat{\beta}_t \sum_{r=0}^{t-1} r^2 a^r = \sum_{r=0}^{t-1} a^r r x_{t-r}$$

It will be seen that the form of these equations is the same as that for the normal equations for the undiscounted case — which it must be, for this corresponds to putting $a = 1$. The sign of $\hat{\beta}_t$ has changed since we are now measuring the time, r, as positive into the past. The other difference is that all the sums are now 'discounted' sums:

$\sum x_i$ is now $\sum a^r x_{t-r}$,

$$t = \sum_{1}^{t} 1 \text{ is now } \sum_{0}^{t-1} a^r,$$

and so on for each term. To get the full value of introducing the weights a, we require evenly spaced data, so we shall assume in what follows that observations occur evenly at times $1, 2, 3, \ldots$. The occurrence of uneven intervals can be dealt with fairly easily once the principles have been grasped.

The forecast of x_{t+h} is, as before,

$$\tilde{x}_{t,h} = \hat{\mu}_t + \hat{\beta}_t h$$

The solution of this forecasting problem looks computationally very difficult, but all the discounted sums required can be calculated very simply by using recurrence relations.

Table 5.6 shows the recurrence relations that can be used for updating the various terms in the normal equations and their application to a set of data. It also indicates the limiting values of the various terms which approach constant values as the amount of data increases.

The calculations used in this method are clearly more involved than those of the previous method discussed. In many circumstances, however, there are advantages in using this method. It does not require the specification of initial conditions and thus can be used with some confidence when only small amounts of data are available.

If we set $a = 1$, the above normal equations become identical with those of the ordinary least squares method. If we have a local constant mean model, corresponding to $\beta =,0$, we obtain the exact exponentially weighted moving average of the previous chapter as the discounted least squares estimate of μ.

Table 5.6 Example of discounted least squares

(a) Recurrence relations for the normal equations

Sum	Symbol	Recurrence relation	Initial value
$\displaystyle\sum_{r=0}^{t-1} a^r$	W_t	$W_t = 1 + aW_{t-1}$	$W_1 = 1$
$\displaystyle\sum_{r=0}^{t-1} ra^r$	A_t	$A_t = a(W_{t-1} + A_{t-1})$	$A_1 = 0$
$\displaystyle\sum_{r=0}^{t-1} r^2 a^r$	B_t	$B_t = A_t + a(A_{t-1} + B_{t-1})$	$B_1 = 0$
$\displaystyle\sum_{r=0}^{t-1} a^r x_{t-r}$	Y_t	$Y_t = x_t + aY_{t-1}$	$Y_1 = x_1$
$\displaystyle\sum_{r=0}^{t-1} ra^r x_{t-r}$	Z_t	$Z_t = a(Y_{t-1} + Z_{t-1})$	$Z_1 = 0$

Normal equations
$$W_t\hat{\mu}_t - A_t\hat{\beta}_t = Y_t$$
$$A_t\hat{\mu}_t - B_t\hat{\beta}_t = Z_t$$

Solutions of normal equations
$$\hat{\mu}_t = C_t(B_t Y_t - A_t Z_t)$$
$$\hat{\beta}_t = C_t(A_t Y_t - W_t Z_t)$$
where
$$C_t = 1/(W_t B_t - A_t^2) - \text{note this is independent of data}$$

Limiting values as $t \to \infty$	$a = 0.8$	$a = 0.9$
$W = 1/(1-a)$	5	10
$A = a/(1-a)^2$	20	90
$B = a(1+a)/(1-a)^3$	180	1440
$1/C = a/(1-a)^4$	500	6300
C	0.002	0.000158

(b) Example, using data from Table 5.4
$a = 0.8$

W_t	A_t	B_t	x_t	Y_t	Z_t	C_t	$\hat{\mu}_t$	$\hat{\beta}_t$	$\hat{x}_{t+2}$
1.000	0.0	0.0	42.8	42.8	0.0	—	—	—	—
1.800	0.800	0.800	42.6	76.84	34.24	1.25	42.6	−0.2	42.200
2.440	2.080	3.328	42.5	103.972	88.864	0.2626	42.487	−0.149	42.189
2.952	3.616	7.942	42.4	125.578	154.269	0.09644	42.386	−0.127	42.132
3.362	5.254	14.501	42.1	142.562	223.877	0.04730	42.146	−0.173	41.800
3.689	6.893	22.697	41.8	155.850	293.151	0.02761	41.874	−0.198	41.478
3.951	8.466	32.138	41.5	166.180	359.201	0.01808	41.578	−0.223	41.132

5.2.3 Double exponential smoothing

In section 4.2 we made use of the recurrence formula for the exponentially weighted moving average. This was

$$\tilde{x}_t = (1 - a)x_t + a\tilde{x}_{t-1}$$

If we carry out ordinary exponential smoothing on the data of Table 5.4, it will be seen that the value of $\tilde{x}_t$ lags behind the trend (see Table 5.7). To find the expected lag, l, we substitute the expected value of x_{t-r}, for the trend model, in the exact form of exponential smoothing. Thus

$$E(\tilde{x}_t) = \Sigma a^r (\mu_t - \beta r)/\Sigma a^r$$

$$= \mu_t - \beta(\Sigma r a^r / \Sigma a^r)$$

$$= \mu - \{a/(1 - a)\}\beta$$

for large t, using the limiting results given in Table 5.5. Thus the lag is $\{a/(1 - a)\}\beta$.

A reasonable way to obtain an estimate of the current mean μ_t is to adjust $\tilde{x}_t$ by adding an estimate of this lag. One way of doing this is to make use of what has been termed double exponential smoothing (see Brown, 1959, 1963). Since the exponentially weighted moving average has the effect of smoothing out much of the fluctuation in the data x_1, $x_2, x_3, \ldots, x_t$, the sequence $\tilde{x}_1, \tilde{x}_2, \tilde{x}_3, \ldots \tilde{x}_t$ is said to be the result of exponentially smoothing the first sequence. Suppose that we now apply the formula of exponential smoothing to $\tilde{x}_1, \tilde{x}_2, \tilde{x}_3, \ldots, \tilde{x}_t$. This we could denote by

$$\tilde{x}_t^{(2)} = (1 - a)\tilde{x}_t + a\tilde{x}_{t-1}^{(2)}$$

The new sequence $\tilde{x}_1^{(2)}, \tilde{x}_2^{(2)}, \tilde{x}_3^{(2)}, \ldots, \tilde{x}_t^{(2)}$ is said to have been obtained by the process of 'double exponential smoothing' or second-order exponential smoothing of the data. If x_t exactly follows the trend line, then $\tilde{x}_t$ lies on a trend line a distance l below, when β is positive. Similarly $\tilde{x}_t^{(2)}$ will be a distance l below $\tilde{x}_t$. Thus we would have

$$x_t = \tilde{x}_t + l$$

$$\tilde{x}_t = \tilde{x}_t^{(2)} + l$$

By subtracting we get

$$x_t = 2\tilde{x}_t - \tilde{x}_t^{(2)}$$

We have here assumed a perfect line. Where x_t is just following a local trend line, then we use the right-hand side as a reasonable estimate of the current mean μ_t. Hence

$$\hat{\mu}_t = 2\tilde{x}_t - \tilde{x}_t^{(2)}$$

To estimate the slope, we use the facts that for the exact trend

$$l = \{a/(1-a)\}\beta$$

and

$$l = \tilde{x}_t - \tilde{x}_t^{(2)}$$

Hence a sensible estimate of β at time t is given by

$$\beta_t = \frac{1-a}{a}(\tilde{x}_t - \tilde{x}_t^{(2)})$$

These are obviously much easier equations to use than the normal equations, though it must be remembered that these are only approximations valid for large t since the lag is only l for large t. For small t these equations can still be used, almost as a separate method, provided we have enough prior information to enable us to put in sensible starting values for $\tilde{x}_0$ and $\tilde{x}_0^{(2)}$. Brown, who developed this method, shows that, if we have μ_0 and β_0 as initial estimates of the mean and slope at the time of the beginning of the data, then

$$\tilde{x}_0 = \mu_0 - \frac{a}{1-a}\beta_0$$

and

$$\tilde{x}_0^{(2)} = \mu_0 - \frac{2a}{1-a}\beta_0$$

give the correct starting values to use. If we can at best make rough guesses for μ_0 and β_0, it is advisable to use the exact method as given by the normal equations, at least to obtain the first few forecasts. Table 5.7 provides an example of the method.

As with Holt's method, we may express the forecasting formulae in terms of the one-step-ahead forecasting errors. The derivation of these formulae is an exercise in algebra which we shall leave to the more mathematically inclined readers. The final equations in error correction form are

$$\hat{\mu}_t = \hat{\mu}_{t-1} + \hat{\beta}_{t-1} + (1-a^2)e_t$$
$$\hat{\beta}_t = \hat{\beta}_{t-1} + (1-a)^2 e_t$$

A comparison of the above error correction forms with those of Holt's method shows that the double smoothing method is just a special case of Holt's method. In fact, if we introduce subscripts h to denote the parameters in Holt's method, then the two sets of equations are identical if we write

$$a_h = a^2$$

Table 5.7 Example of double exponential smoothing

(a) Formulae

$$\tilde{x}_t = (1-a)x_t + a\tilde{x}_{t-1}$$

$$\tilde{x}_t^{(2)} = (1-a)\tilde{x}_t + a\tilde{x}_{t-1}^{(2)}$$

$$\hat{\mu}_t = 2\tilde{x}_t - \tilde{x}_t^{(2)}$$

$$\hat{\beta}_t = \frac{1-a}{a}(\tilde{x}_t - \tilde{x}_t^{(2)})$$

$a = 0.8, \hat{\mu}_0 = 42.9$, $\hat{\beta}_0 = -0.15$, assumed initial values giving

$$\tilde{x}_0 = 43.5, \tilde{x}_0^{(2)} = 44.1$$

(b) Data from Table 5.4

x_t	$\tilde{x}_t$	$\tilde{x}_t^{(2)}$	$\hat{\mu}_t$	$\hat{\beta}_t$	$\hat{x}_{t+2}$
	43.5	44.1	42.9	−0.15	−
42.8	43.32	43.944	42.696	−0.156	42.384
42.6	43.176	43.790	42.562	−0.153	42.256
42.5	43.041	43.641	42.441	−0.150	42.141
42.4	42.913	43.495	42.331	−0.145	42.041
42.1	42.750	43.346	42.154	−0.149	41.856
41.8	42.560	43.189	41.931	−0.157	41.617
41.5	42.348	43.021	41.675	−0.168	41.339

and

$$c_h = (1-a)^2$$

Thus if in Holt's method the parameters are related by

$$c_h = (1 - \sqrt{a_h})^2$$

the method is identical to the double smoothing method.

Double exponential smoothing is also related to the method of section 5.2.2, since it can be shown that for large t the solutions of the normal equations are equivalent to the estimates of μ_t and β_t given by the use of double exponential smoothing.

5.2.4 Conclusions

We have now considered three methods of estimating the parameters and forecasting for a local linear trend model. Table 5.8 gives a summary showing the relations between them. The practical decision as to which method to use requires the choice of the best forecasting parameters (a or a and b), which is discussed in section 15.3, and the comparison of the methods on real or suitably simulated data.

Table 5.8 Summary of methods of section 5.2

	Holt's method	Discounted least squares (DLS)	Double exponential smoothing
Basis of derivation	Construction of recurrence relations based on model structure	Estimation of model parameters using method of DLS	Construction of unbiased estimators out of single and double exponentially smoothed data, assuming t large
Number of forecasting parameters	2	1	1
Initial estimates of μ_t and β_t required	Yes	No	Yes
Relations between methods		⟶	A special case of Holt's method
		⟶	Equivalent to DLS for large t

The reason for discussing all three methods at some length was not just to cover the main methods. The three methods we have discussed illustrate three main approaches to forecasting local models, which are

(a) to use discounted versions of classical statistical methods (section 5.2.1).
(b) to develop intuitively recurrence relations for the parameters (section 5.2.2).
(c) to use a simple method and then adjust for the bias, or lag (section 5.2.3).

We will come across all these methods again in later chapters.

5.3 Polynomial models

An obvious extension of the trend model is to move to polynomial models involving t^2, t^3, etc., to describe more elaborate movements in the data. All the methods described in this chapter have natural extensions that provide forecasts for such situations.

A word of caution is, however, advisable in relation to using polynomial models. As a rule quadratic trends and higher order polynomials should only be used if there is some logical basis for such a structure. In general the magnitudes of errors both from incorrect choice of model and from changes in structure in time increase with the complexity of the model used. Further, the more parameters used the more observations are needed to obtain forecasts. Even with global models a lot of data are required to give a reasonable error variance. Cowden (1963) tabulates the type of information required to assess the error variances. By way of brief illustration, if, as usual, the ϵ values are independent with variance σ^2, then for, say, eleven observations the one-step-ahead forecast error variance is 1.647 σ^2 for a linear trend model, 2.098 σ^2 for a quadratic model, 2.771 σ^2 for a cubic model and 3.908 σ^2 for a quartic model. Clearly, in a less stable situation with a model fitted by, say, discounted least squares the error variances will be even higher in practice.

References

Brenner, J. L., D'Esopo, D. A., and Fowler, A. G. (1968). Difference equations in forecasting formulae. *Man. Sci.*, 15, No. 3.

Brown, R. G. (1959). *Statistical Forecasting for Inventory Control.* McGraw-Hill, New York.

Brown, R. G. (1963). *Smoothing, Forecasting and Prediction of Discrete Time Series.* Prentice-Hall, Englewood Cliffs, New Jersey.

Coutie, G. A., Davis, O. L., Hassall, C. H., Miller, D. W. G. P., and Morrell, A. J. H. (1966). *Short Term Forecasting.* ICI Monograph No. 2. Oliver and Boyd., Edinburgh.

Cowden, D. J. (1963). The perils of polynomials. *Man. Sci.*, 9, No. 4, 546—550.

Croxton, F. E., Cowden, D. J., and Klein, S. (1968). *Applied General Statistics.* Pitman.

Gilchrist, W. G. (1967). Methods of estimation involving discounting. *J. Roy. Statist. Soc.* (B), 29, 355—369.

Holt, C. C. (1957). *Forecasting Seasonals and Trends by Exponentially Weighted Moving Averages.* Carnegie Institute of Technology, Pittsburgh, Pennsylvania.

Chapter 6

Regression models

6.1 Introduction

In the previous chapters we have considered methods of forecasting that only make use of past values of the variable being forecast. Clearly, this approach is limited as it is reasonable that to obtain the best forecasts we may need to make use of information from a wide variety of sources. In particular, we need to make use of the information contained in other variables that are related to the one being forecast. For example, in forecasting the sales of one particular product we may relate the sales S to the wealth in the consumer's pocket, as measured by some income variable I, the price of the product P, the expenditure on advertising A, and probably other variables as well. We would thus be looking for a model of the general form

$$S = f(I,P,A, \ldots)$$

where $f(\ )$ is some function to be found. Our ability to construct such models that provide good forecasts is a measure of our real understanding of the situation. Clearly, in constructing models for this type of forecasting a good knowledge of the underlying situation is essential. There are a number of books that deal particularly with this area of forecasting, e.g. those by Robinson (1971) and Spencer, Clark and Hoguet (1961).

Within the scope of this book we are concerned only with some of the statistical aspects of the use of such models. There are an infinity of possible models that might be used, so we must clearly limit our discussion to a simple type as an introduction to the subject. In this chapter we will consider the problem of forecasting with the most commonly used model, namely that based on the use of linear relationships. By way of example, suppose that in the situation described above we finally obtained the model

$$S_t = 127.12 + 2.34\,I_t - 0.32\,P_t + 0.65\,A_{t-1} + 6.82\,t + \epsilon_t$$

This model is linear in its parameters. It will be seen that S_t depends not only on the current values of I and P but on the previous value of A and on an underlying trend represented by the 6.82 t. As usual, the random variable ϵ_t represents the failure of the world to behave in the nice regular fashion that would make forecasting so much easier. To consider the forecasting use of such models, let us limit the consideration to the simplest case where the model relates a variable y_t to just one other variable, x_t. The linear model would thus be

$$y_t = \alpha + \beta x_t + \epsilon_t$$

If we can use the data to estimate the unknown parameters, α and β, given estimators a and b, then we have as a fitted model the line

$$y = a + bx$$

These forms of model are called linear regression models and we talk about fitting a regression line of y on x. If we wish to forecast y_{t+h}, the natural thing to do is to use

$$\hat{y}_{t+h} = a + bx_{t+h}$$

This is fine, provided that we know the future value x_{t+h}. There are situations where this is the case, as, for example, where x_{t+h} is the number of selling points of a product, in which case x_{t+h} is a value under our control. In general, however, x_{t+h} may be as much a quantity requiring forecasting as y_{t+h}. If $\hat{x}_{t+h}$ is a forecast of x_{t+h}, then

$$\hat{y}_{t+h} = a + b\hat{x}_{t+h}$$

will provide a forecast of y_{t+h}. This forecast may be no better than that obtained by looking at past values of y_t and using, say, a local trend model. It does, however, provide another forecast based on some additional information. This may be usefully combined with the trend forecast using the methods to be discussed in Chapter 17.

It is sometimes the case that past experience shows that $\hat{x}_{t+h}$ leads to a better forecast of the y variable than we could obtain from past values of the y variable. In this case, using the regression model provides a good way of forecasting y_{t+h}. As an alternative procedure worth considering, we could examine as a basic model

$$y_t = \alpha + \beta \hat{x}_t + \epsilon_t$$

where $\hat{x}_t$ is the forecast of x_t. We would then fit this to past values of y and $\hat{x}$ to obtain a fitted line

$$y = a' + b'\hat{x}$$

and forecast y_{t+h} by

$$\hat{y}_{t+h} = a' + b'\hat{x}_{t+h}$$

In a different type of situation a change in a variable x may not produce an effect on y for some time, h. For example, sales of spares y are likely to lag behind sales x of the item for which they are spares. In such a situation we might have a regression model of the form:

$$y_t = \alpha + \beta x_{t-h} + \epsilon_t$$

where x_{t-h} is referred to as a lagged variable. We can now forecast y_{t+h} by fitting and extrapolating to get

$$\hat{y}_{t+h} = a'' + b'' x_t$$

where x_t is the latest known value of x.

Clearly, it is more common to have the variable y depending on more than one other variable, e.g. ice cream sales may depend on both temperature and sunshine. If we denote these variables by $x_{1,t}$, $x_{2,t}, \ldots, x_{k,t}$, then we may write our general model as

$$y_t = \sum_{i=0}^{k} \beta_i x_{i,t} + \epsilon_t$$

The constant term is allowed for by simply letting $x_{0,t} = 1$ for all t; then β_0 is the constant. We will use this form as a general notation; the $x_{i,t}$ may represent an observation at time t, a lagged observation at some previous time, a forecast of a variable at time t or possibly a function of time. These $x_{i,t}$ values, which we will term the regressor variables, are regarded as a set of given constant values. The variable y_t, the dependent variable, is thus composed of a deterministic part, $\Sigma \beta_i x_{i,t}$, and an added random variation, ϵ_t.

Before turning to some of the details of fitting and forecasting regression models, it is important to emphasize that our ultimate concern is in forecasting. Much of the statistics concerned with regression deals with measuring how good the model was and how well past y were explained by the right-hand side of the regression equation. It can easily happen that a very good model, looked at in purely statistical terms, can lead to very poor forecasts because of problems of reliability of data and of inability to get good forecasts of regressor variables. Our ultimate criteria of success cannot be the various measures of fit described in textbooks dealing with regression; our criteria must be criteria such as the mean square forecasting error that refer directly to the forecasting use that we make of the regression model.

6.2 Fitting regression models

We have discussed the form of regression models used for forecasting. We now turn to examine briefly the question of fitting these models.

There is a great deal of literature on this topic (see, for example, Draper and Smith, 1966). In nearly all this literature the models are regarded as being, in theory, global models, though in practice the users of the models often ignore data from the far past as being irrelevant. From the forecasting viewpoint it seems more reasonable to assume that the models are best treated as local models. Thus rather than use least squares we would use discounted least squares with a discounting factor a^r. If it so happens that we are wrong here, then in choosing a to get the best forecasts we find, approximately, that $a = 1$. We would thus obtain the same forecasts as the user of ordinary least squares fitting.

To indicate the type of result we obtain by applying discounted least squares, consider the data in terms of the fitted model

$$y_t = \sum_{i=0}^{k} b_i x_{i,t} + \hat{e}_t$$

where $\hat{e}_t$ is the error in the fitted regression, the residual. The regression coefficients b_i are the discounted least squares coefficients when they are chosen to minimize

$$S = \sum_{r=0}^{t-1} a^r \hat{e}_{t-r}^2$$

Substituting for $\hat{e}$ and minimizing with respect to b_0, putting $x_{0,t} = 1$, leads to

$$\sum_{r=0}^{t-1} a^r \left(y_{t-r} - \sum_{t=0}^{k} b_i x_{i,t-r} \right) = 0$$

Minimizing with respect to any b_i gives

$$\sum_{r=0}^{t-1} a^r x_{j,t-r} \left(y_{t-r} - \sum_{i=0}^{k} b_i x_{i,t-j} \right) = 0$$

(see Appendix D). If we denote,

$$\sum_{r=0}^{t-1} a^r z_{t-r} v_{t-r}$$

by $S(z, v)$, the above equations giving the b values, are

$$b_0 S(1,1) + b_1 S(1, x_1) + \ldots + b_k S(1, x_k) = S(1, y)$$

$$b_0 S(x_1, 1) + b_1 S(x_1, x_1) + \ldots + b_k S(x_1, x_k) = S(x_1, y)$$

$$b_0 S(x_k, 1) + b_1 S(x_k, x_1) + \ldots + b_k S(x_k, x_k) = S(x_k, y)$$

These equations, the normal equations, can be compared with the equation of the fitted regression

$$b_0 + b_1 x_1 + \ldots + b_k x_k = y$$

Thus the $k+1$ normal equations are obtained by multiplying this equation by a^r, $a^r x_1, \ldots, a^r x_k$, respectively, and summing each over time. Solving these equations gives the coefficients $b_0, \ldots, b_k$. Table 6.1 gives an example of such a calculation. In (a) the ordinary least squares line is fitted. Clearly, in the calculation the order in which the observations occurred is irrelevant. In the normal equations $S(u,v)$ represents the sum of products of u and v. To calculate the normal equations in the discounted case, we note that if $S_t(u,v)$ denotes the discounted sum at time t then

$$S_t(u,v) = u_t v_t + a S_{t-1}(u,v)$$

where $S_1(u,v) = u_1 v_1$. Therefore we can use the usual type of recurrence relation to calculate these quantities and thus write down the normal equations at any time t. The normal equations of Table 6.1(b) are those at time 12. Figure 6.1 shows the data and the fitted lines for $a = 1.0$, $a = 0.95$ and $a = 0.90$. It will be seen that as a decreases the line moves closer to the most recent data points, as one would wish it to for forecasting purposes.

Almost all computers and the more sophisticated desk-top calculators provide programmes that solve the normal equations and estimate the regression coefficients b_1 using ordinary least squares. These can be used to obtain the discounted least squares solutions by discounting all the regressor variables $x_{i,t-r}$ (including $x_{0,t-r}$) by multiplying each by the value of $(\sqrt{a})^r$. The logic of this is shown by the following reformulation of the discounted least squares criterion:

$$\Sigma a^r \epsilon_{t-r}^2 = \Sigma [(\sqrt{a})^r y_{t-r} - b_0 (\sqrt{a})^r x_{0,t-r} - \ldots - b_k (\sqrt{a})^r x_{k,t-r}]^2$$

In using forecasts from a regression model we will wish to develop some idea of the error variance. Let us look briefly at this for the simplest model

$$y_t = \beta x_t + \epsilon_t$$

If x_{t+h} is a known quantity, then

$$\hat{y}_{t+h} = b x_{t+h}$$

and

$$\begin{aligned} \text{Var}(e_{t+h}) &= \text{Var}(y_{t+h} - \hat{y}_{t+h}) \\ &= \text{Var}(\beta x_{t+h} + \epsilon_{t+h} - b x_{t+h}) \\ &= \text{Var}(b) x_{t+h}^2 + \sigma^2 \end{aligned}$$

Notice that in this simple case the variance of the error increases as the

Table 6.1 Fitting a regression line

(a) Classical calculation

t	y	x	xy	x^2	y^2
1	1.2	28.3	33.96	800.89	1.44
2	1.3	26.8	34.84	718.24	1.69
3	1.3	28.7	37.31	823.69	1.69
4	1.6	35.1	56.16	1232.01	2.56
5	1.6	37.0	59.2	1369.00	2.56
6	1.4	30.5	42.7	930.25	1.96
7	1.5	32.5	48.75	1056.25	2.25
8	1.6	35.7	57.12	1274.49	2.56
9	1.5	34.4	51.6	1183.36	2.25
10	1.7	37.3	63.41	1391.29	2.89
11	2.0	39.2	78.4	1536.64	4.00
12	2.1	43.2	90.72	1866.24	4.41
Totals	18.8	408.7	654.17	14182.35	30.26

Normal equations, $a = 1$

$b_0 S(1, 1) + b_1 S(1, x) = S(1, y) = 12 \, b_0 + 408.7 \, b_1 = 18.8$

$b_0 S(1, x) + b_1 S(x, x) = S(x, y) = 408.76 + 14182.35 \, b_1 = 654.17$

$$b_0 = -0.232$$
$$b_1 = 0.053$$

(b) Discounted regression, based on table of (a), using recurrence form
$$S_t(u, v) = uv + aS_{t-1}(u, v)$$

t	$\Sigma a^r y_{t-r}$	$\Sigma a^r x_{t-r}$	$\Sigma a^r x_{t-r} y_{t-r}$	$\Sigma a^r x_{t-r}^2$	$\Sigma a^r y_{t-r}^2$	Σa^r
1	1.2	28.3	33.96	800.89	1.44	1
2	2.44	53.68	67.10	1479.09	3.06	1.95
3	3.62	79.70	101.06	2228.83	4.60	2.85
4	5.04	110.82	152.16	3349.40	6.93	3.71
5	6.38	142.28	203.75	4550.93	9.14	4.52
6	7.47	165.66	236.27	5253.63	10.64	5.30
7	8.59	189.88	273.20	6047.20	12.36	6.03
8	9.76	216.08	316.66	7019.33	14.30	6.73
9	10.77	239.68	352.43	7851.72	15.84	7.39
10	11.94	265.00	398.22	8850.43	17.94	8.03
11	13.34	290.95	456.71	9944.55	21.04	8.62
12	14.77	319.60	524.59	11313.56	24.40	9.19

Normal equations, $a = 0.95$ and $t = 12$

$9.19 \, b_0 + 319.60 \, b_1 = 14.77$

$319.60 \, b_0 + 11313.56 \, b_1 = 524.59$

$$b_0 = -0.284$$
$$b_1 = 0.054$$

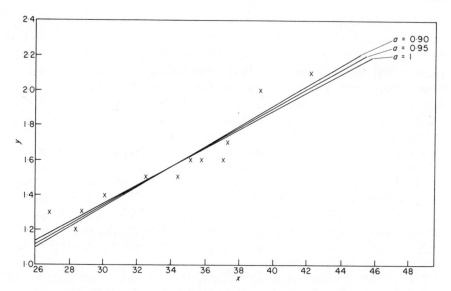

Figure 6.1   Plot of the data and regression lines of Table 6.1

square of the distance of x_{t+h} from the origin, which acts as a pivot. In the general case the exponentially weighted mean of the x values acts as the pivot. It is the uncertainty in the estimate of slope that generates the element $\text{Var}(b)x_{t+h}^2$ in the error variance.

If x_{t+h} has to be forecast by $\hat{x}_{t+h}$, then

$$\text{Var}(e_{t+h}) = \text{Var}(\beta x_{t+h} + \epsilon_{t+h} - b\tilde{x}_{t+h})$$
$$= \text{Var}(b\tilde{x}_{t+h}) + \sigma_\epsilon^2$$
$$= E(b^2)E(\hat{x}_{t+h}^2) - \beta^2 x_{t+h}^2 + \sigma_\epsilon^2$$
$$= \{\text{Var}(b) + \beta^2\}\{\text{Var}(\hat{x}_{t+h}) + x_{t+h}^2\} - \beta^2 x_{t+h}^2 + \sigma_\epsilon^2$$
$$= \text{Var}(b)x_{t+h}^2 + \sigma_\epsilon^2 + \text{Var}(\hat{x}_{t+h})\{\text{Var}(b) + \beta^2\}$$

assuming that b and $\hat{x}_{t+h}$ are independent and $\hat{x}_{t+h}$ is an unbiased forecast of x_{t+h}. Thus the expected mean square error is increased by an additional amount depending not only on $\text{Var}(\hat{x}_{t+h})$ but also on $\text{Var}(b)$ and β^2. As usual, we need to investigate the magnitude of the mean square error directly via a study of the observed errors when the method is applied to past data. Theil (1966) gives a general study of forecast errors in regression models.

6.3 Selecting the variables

In choosing the variables to use in regression for forecasting one must of necessity start by using one's knowledge of the situation to pick likely candidates. Very often, however, there are many possible candidates and it is a real problem to choose a small number to use for one's actual forecasting exercise. The need for a small number of regressor variables is not at first obvious. If we only have a few observations and a comparable number of variables, then the fitted regression could be made to fit very closely to the data. However, this would be picking up the peculiarities of the individual observations and the variances of the b values would be high. The variance of $\hat{y}_{t+h}$ would therefore be unacceptably large. To get reasonable forecasts we therefore need many more observations than variables. This in practice, where data is often short, puts a constraint on the number of variables used. A further aspect is that in many practical situations most of the variability in y can be explained in terms of a few of the variables. Though the *fit* can always be improved by introducing new variables, the amount of the improvement may be small and the quality of the *forecasts* may well deteriorate.

The problem thus arises as to how to choose a small number of regressor variables from the original set chosen. There are a large number of different ways of doing this. e.g. see Allen (1971) and Beale (1970).

In common practice most methods assume a global model and are based on:

(a) repeated examinations of the correlation of the fitted regression and the dependent variable y;
(b) studying the changes in this produced by adding or removing the various regressor variables to and from the model;
(c) studying the significance of the regression coefficients obtained.

They are thus based on the use of statistics such as the correlation coefficient and t statistic.

The amount of calculation involved in these methods is very large and a computer is really essential. All major computers have programmes available that can be used in the selection of variables.

From the forecasting viewpoint the assumption of a global model as the basis for selecting variables is not so reasonable. It is clearly advisable to select the variables that provide the best explanation of the most recent data rather than the averaged-out sort of selection made by the above methods. If we admit that our real data may not be from a stable, global, situation, but rather that a local approximate model-building approach is more appropriate, then two things follow. Firstly, the assumptions underlying the above methods will not be valid. Secondly, it would intuitively appear both reasonable and possible to replace the correlation coefficients, t, statistics, etc., of the above methods by modified versions based on discounted statistics. These can then be used in modified versions of the above methods. A detailed discussion of such an approach is given by Singleton (1971). By way of a simple illustration, consider the multiple correlation coefficient R, defined by

$$R^2 = 1 - \sum_{i=1}^{t} \hat{e}_i^2 \bigg/ \sum_{i=1}^{t} (y_i - \bar{y})^2$$

where

$$\bar{y} = \sum_{i=1}^{t} y_i / t$$

This is the simplest measure of the ability of our model to 'explain' the data. If our model fitted perfectly, the $\hat{e}$ values, the residuals, would all be zero so R^2 would be one. If it did not really 'explain' the data at all, the variability of the $\hat{e}$ values would be much the same as that of the y values. In this case $\Sigma \hat{e}^2$ would be approximately the same as $\Sigma(y_i - \bar{y})^2$, so that R^2 would be approximately zero. Most of the methods above make some use of the magnitude of R^2 in selecting the regressor variables. If we now say that interest is concentrated on the most recent information, then it is reasonable to introduce discounting into the above statistics. Thus if our usual exponential weights are used, we would replace

$$\sum_{i=0}^{t} \hat{e}_i^2$$

by

$$\sum_{r=0}^{t} a^r \epsilon_{t-r}^2$$

and

$$\sum_{i=0}^{t} (y_i - \bar{y})^2$$

by

$$\sum_{r=0}^{t} a^r (y_{t-r} - \bar{y})^2,$$

where $\bar{y}$ may be the ordinary mean or the exponentially weighted mean. We would thus define a discounted multiple regression coefficient by

$$R_d^2 = 1 - \sum_{r=0}^{t} a^r \hat{e}_{t-r}^2 \bigg/ \sum_{r=0}^{t} a^r (y_{t-r} - \bar{y})^2$$

Table 6.2 Some illustrative values
of R_d^2

Discounting factor a	Variables $(3, 5, 6)$ R_d^2	Variables $(4, 5, 6)$ R_d^2
1.00	0.83	0.81
0.99	0.90	0.89
0.98	0.94	0.94
0.97	0.96	0.96
0.96	0.96	0.97

The properties of R_d^2 of relevance to the selection of variables are essentially the same as those of R^2, so it can be used to form the basis of methods for the selection of variables that are most important locally, which is the object in forecasting situations.

Table 6.2 shows the values of R_d^2 from an exercise in which there were five regressor variables (labelled 2, 3, . . . , 6). The values of R_d^2 are used to compare the effectiveness of the sets (3,5,6) and (4,5,6). It is seen that for $a = 1.00$, i.e. for the usual multiple correlation coefficient, the first set is better than the second. However, if we put more emphasis on recent data and set $a = 0.96$, then we begin to get the opposite picture.

We have referred to the use of lagged variables as regressors. In practice one is not often sure as to the length of the lag. In initial explorations one therefore needs to put in a range of different lags as possible regressor variables. It is then up to the regression program to select the most appropriate ones. The selection of variables in this fashion gives the selection of regressors for obtaining the best fit or the best forecasts over past data. It does not necessarily pick the variables that are clear causal factors in determining the values of the independent variable. It is then up to the forecaster to decide whether the results obtained are chance results, depending on the chance form of the data used, or whether they represent the real effects necessary to obtain good forecasts in the future. To do this the forecaster, as always, needs to know a great deal about the variables in the forecasting situation. The use of the programs for selecting variables is thus part of the preliminary exploration of the data, and the forecaster should use a wide range of regressor variables in this exploration.

It may be helpful at this stage to list some of the types of variables that have proved useful in forecasting situations:

(a) Forecasts of a variable within a firm, say its sales, often use various measures derived at a Government level. As indicated before, these might be actual values, forecast values or lagged values. A list of sources of American economic and other variables of use in forecasting are given in Woy (1965). A large number of business and trade journals also publish relevant data of value for particular industries.

(b) It is often of value to include functions of time in the regression model. Thus we might have

$$y_t = \beta_0 + \beta_1 t + \beta_2 x_{2,t} + \epsilon_t$$

A seasonal variation can be modelled, using weekly data, by using harmonic terms. For example,

$$y_t = \beta_0 + \beta_1 \cos 2\pi n/52 + \beta_2 \sin 2\pi n/52 + \beta_3 x_{3,t} + \epsilon_t$$

where $n = 1, 2, \ldots, 52$ is the week number in the year.

(c) A rather similar device is sometimes used to deal with quarterly or monthly seasonals or the effects on sales of holidays, etc. Here we define a series of x variables that are zero everywhere except at the month or holiday week in question. The value 1 is given at that time. The term βx is only non-zero at that time and then its estimated value $b.1$ gives the magnitude of the seasonal or holiday effect. If there are monthly seasonals or a large number of holidays, we have a corresponding number of parameters whose values therefore cannot be estimated as precisely as when only a small number were used. One way round this, as a fairly rough way out, is to use past data to invent a function whose shape seems about right for the relative magnitudes of the various effects. For example, it might be that the holiday effect is: 1 in weeks 12, 19 and 35, 2 in weeks 3, 30 and 48, and zero elsewhere. The coefficient, when estimated, will then give an improved estimate of the level of the holiday effects, though obviously, in the example, it would not alter the 2:1 ratio chosen rather arbitrarily. Thus if the value of b was 0.7, as happened in a particular case, the estimated holiday effects become 0.7 and 1.4 on the appropriate weeks.

(d) Considerable thought must be given to the way in which any regressor variable will influence the dependent variable. It is sometimes the case that it is only the general level of a particular regressor that is important. The random variation in this variable is thus an irrelevance that will make our forecasts worse. An obvious way of dealing with this is to try and smooth out this random variable. The smoothed data would then be used as the regressor variable. By way of a further example, in a sales forecasting exercise, the effect of temperature was thought to be important. However, on reflection an important feature was not the actual temperature but whether that temperature was greater or less than the norm expected by people at that time of the year. Thus two series were used; one, T_t, was the seasonal normal temperature, the other, ΔT_t, was the deviation of the actual temperature from this at time t. In the event it was found that T_t was effectively allowed for by the seasonal terms in the model and ΔT_t remained as a highly significant term.

(e) Suppose that variables $x_{1,t}$ and $x_{2,t}$ are thought to be relevant for forecasting the y_t series. It does not necessarily follow that x_1 and x_2 should appear in the regression model as

$$y_t = \beta_1 x_{1,t} + \beta_2 x_{2,t} + \epsilon_t$$

It could be, for example, that the best model for forecasting is

$$y_t = \beta_1 x_{1,t}^2 + \beta_2 \log x_{2,t} + \beta_3 x_{1,t} x_{2,t} + \epsilon_t$$

In other words, the ways in which $x_{1,t}$ and $x_{2,t}$ influence y_t are not necessarily simple or linear. Note, however, that the model is still linear in the parameters and is still straightforward to fit. Programmes to select variables can be used to look at the use of functions of regressor variables, the functions being fed in as though they were totally separate regressors. Care, as usual, must be taken in the interpretation of the outcomes of such exercises, but if the forecaster obtains a model that gives better forecasts then the exercise is worthwhile.

6.4 Econometric models

As the title of this book is *Statistical Forecasting* the writer does not wish to devote much space to economic forecasting. There are a number of good books in this area (e.g. Theil, 1966) and to do the topic any justice at all would require far more space than is available. There are, however, many statistical aspects of economic forecasting and, indeed, some books of econometrics could well have the same title as this chapter. Again, however, much of the statistics involved arise out of the nature of the economic theory used and will not be dealt with here. The aim of this section is to show briefly how the subject links in with what we have said so far. The reader may then know whether in his problems he needs to move on to a study of econometric models.

The first and obvious comment about forecasting with models using economic variables is that to understand and use such models properly a knowledge of economics is necessary. At the very least, the advice of a capable economist should be obtained, for there are many pitfalls for the unwary in economic forecasting of any sort.

It is useful in considering models involving economic variables, econometric models, to classify the variables. The usual terminology defines:

(a) endogenous variables — variables occurring in the equations of the model whose behaviour is 'explained' by the model itself;
(b) exogenous variables — variables external to the situation that effect it but are not affected by it;
(c) predetermined variables — mathematical terms such as trends;
(d) lagged variables — past values of variables (a) and (b).

By way of example, suppose we try to fit a model that relates our firm's sales to our prices, one or two indexes of income of our buyers and their economic state and possibly include a trend term in the model. We could also relate our prices to past demand, etc. Of these variables our sales and prices would be endogenous variables and the various indexes would be exogenous variables, as they would not in general be affected by anything in our situation. The trend would be a predetermined variable.

A problem of most econometric models is that we usually wish to forecast the endogenous variables, but yet the two types of variables tend to be mixed up together in rather involved regression models. For example, if A and B are exogenous and x and y endogenous, we might have

$$y_t = \beta_0 + \beta_1 x_{t-1} + \beta_2 y_{t-1} + \epsilon_t$$
$$x_t = y_t + A_t + B_t$$

Such equations do not lend themselves to simple direct solution by least squares. Much of the literature of econometric models is concerned with the least squares solution of models of such involved types. We will not study this topic further here other than to give, in Figure 6.2, a visual presentation of such equations that is of practical use. Assuming that the parameters are estimated, we still have the quantities A_t and B_t to deal with. In some situations we will be asking

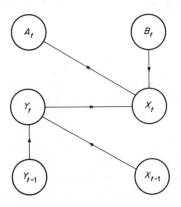

Figure 6.2   An arrow scheme

questions about what x_{t+1} will be if A_{t+1} and B_{t+1} take on various different values. These would be referred to as 'conditional' forecasts. If, however, we have to use some forecasts of A_{t+1} and B_{t+1} to obtain our forecast of x_{t+1}, we would have an 'unconditional' forecast. A practical aspect of most econometric models is the large number of equations involved. The paper by Ball and Burns (1968) gives an example of such a set of equations for a relatively small model, together with a discussion of the forecasting aspects of such models.

A particular problem occurs in economic forecasting, and in general in forecasting with regression models, that the forecaster needs to be aware of. The problem is usually referred to as the problem of multicolinearity. In essence the problem is this: if in the model

$$y_t = \beta_0 + \beta_1 x_{1t} + \beta_2 x_{2t} + \epsilon_t$$

it so happens that $x_{2t} = \gamma x_{1t}$, for all t, γ being a constant, then the normal equations will not give a solution. The reason is clear if we substitute to get

$$y_t = \beta_0 + \eta x_{1t} + \epsilon_t$$

where $\eta = \beta_1 + \gamma \beta_2$. The equation for y_t really only has two parameters, but yet we were originally trying to solve it for three. In practice we will not get such an exact relation between the two variables, but we do often get very high correlations. Though in these cases we can solve the normal equations, the variances of the estimators we get can become unacceptably high. One way out of this dilemma is to try and replace x_1 and x_2 by a new single variable that combines their effects. For example, in a regression relating sales to temperature and sunshine a variable termed 'weather' could be introduced which combined both factors in an acceptable fashion. It is never very clear in practice just how large will be the errors caused by multicolinearity effects. The safest approach is to use the fitted regression model on new data and examine the forecast errors.

6.5 Time as a regressor variable

The general changes in variables showing as trends or seasonal variations have sometimes been regarded as a great nuisance in regression analysis and only data adjusted to remove these effects have been used in regression. However, the inclusion of trends and harmonic terms such as illustrated in forecasting situation (b) of section 6.3 is a perfectly acceptable device. If the other regressor variables do not show any such trend or seasonal, the interpretation of the terms will be quite clear. A problem does arise, however, when a regressor variable also follows its own trend or seasonal pattern. Suppose, for example, that it follows a trend with a number of clear divergences from the trend which influence the dependent variable. When the model is fitted by least squares or discounted least squares, the coefficient of this regressor variable will adjust to give the best explanation of the effects of these divergences on the dependent variable. However, this coefficient will also ensure that the trend in the regressor variable will contribute to the trend in the dependent variable. The term on the right-hand side put in to allow for trend will thus only need to account for the additional trend required to get the appropriate magnitude of trend for the dependent variable. For example, consider the model

$$y_t = \alpha + \beta t + \gamma x_t + \epsilon_t$$

Suppose x_t has a trend of about 1 unit/unit of t, together with deviations that are reduced by a factor of ½ in their influence on y_t. Suppose that y_t has a trend of about 2 units/unit of t. In fitting this model γ will have a value of about ½ to get the effect of the deviations right. This will thus contribute a trend of ½, so β will have to be about $\frac{3}{2}$ to get the slope of 2 correct. Thus though we will get a good fit, we will not get a fitted model with a clearly interpretable trend term.

An alternative is to use not x_t but a detrended series in which we have fitted a trend to x_t and then subtracted it, leaving only the deviations from trend. The inclusion of x_t means that the estimates of the trend components may not be so good as if the trend were fitted on its own. This suggests that, if the time function is a dominant feature of the data and the other regressor variables, economic variables, say, do not show this feature strongly, then a reasonable approach is to fit the time function first. The residuals of this, which in the above are the 'detrended' series, are then regressed on the other regressor variables. Thus, for example, in the above we would first fit

$$y_t = \alpha + \beta t + \eta_t$$

and obtain the residuals

$$\hat{\eta}_t = y_t - \hat{\alpha} - \hat{\beta} t$$

These would then be fitted to x_t, or to a similarly detrended x_t, by

$$\hat{\eta}_t = \gamma x_t + \epsilon_t$$

giving $\hat{\gamma}$. The final fitted model would be

$$y_t = \hat{\alpha} + \hat{\beta} t + \hat{\gamma} x_t$$

This fitted model does not have the same estimated parameters as would have been obtained from applying least squares to the original model. This method, called stagewise least squares, is discussed by Draper and Smith (1966) and, in a forecasting context, by Spencer, Clarke and Haguet (1961).

Following the remarks made in section 6.3(e) on the possible use of more involved functions, it is perhaps worth noting that many processes do show quite involved variation in time. It is well worthwhile considering which are the best ways of modelling such behaviour. The method of characteristic modes discussed in Farmer (1964) and Ivakhnenko and Lapa (1967) provides an approach to choosing and using appropriate functions of time for constructing forecasting models. A treatment of this theory is beyond the scope of this book, but it should be noted that it emphasizes the value of using linear models, so that we have a model of the general form:

$$y_t = \alpha\phi_1(t) + \beta\phi_2(t) + \epsilon_1$$

where $\phi_i(t)$, $i = 1, 2, \ldots$, are appropriate functions forming the natural 'modes' of change of the variable. It also brings out the value of choosing functions for $\phi_1(t)$, $\phi_2(t)$, etc., that are totally distinct in behaviour (in theoretical terms — orthogonal). There is no point in having a feature in the shape of $\phi_2(t)$ which is already in $\phi_1(t)$.

This brief discussion of time as a regressor variable will be developed further in the chapter on seasonal forecasting (Chapter 8).

References

Allen, D. M. (1971). Mean square error of prediction as a criteria for selecting variables. *Technometrics*, 13, 469—475.

Ball, R. J. and Burns, T. (1968). An econometric approach to short run analysis of UK economy 1965—66. *Op. Res. Quarterly*, 19, 225—256.

Beale, E. M. L. (1970). Note on procedures of variable selection in multiple regression. *Technometrics*, 12, 909—914.

Draper, N. R. and Smith, H. (1966). *Applied Regression Analysis*. John Wiley and Sons. New York.

Farmer, E. D. (1964). A method of prediction for non-stationary processes. *Proceedings of Second Congress of the Intern. Fed. Auto Control*. Butterworths. London.

Ivakhnenko, A. G. and Lapa, V. G. (1967). *Cybernetics and Forecasting Techniques*. Elsevier. New York.

Robinson, C. (1971). *Business Forecasting*. Thomas Nelson and Sons, London.

Singleton, P. (1971). *Multiple Regression Methods in Forecasting*. Thesis M.Phil., London University.

Spencer, M. H., Clark, C. and Hoguet, P. W. (1961). *Business and Economic Forecasting in Econometric Approach*. Irwin, Homewood, Illinois.

Theil, H. (1965). *Applied Economic Forecasting*. North-Holland Publishing Co., Amsterdam.

Woy, J. B. (1965). *Business Trends and Forecasting — Information Sources*. Gale Research Co., Detroit.

Chapter 7

Stochastic models

7.1 Introduction

We consider in this chapter the use of stochastic models in forecasting, that is to say, the use of models in which the random element plays the dominant part in determining the structure of the model. In previous models the chance element ϵ_t was simply an added 'error' added separately at each time moment to a strictly deterministic function. Appendix A brings together some of the terminology and definitions of stochastic theory as relevant to forecasting. The reader should, however, be able to follow most of the results of this chapter at an intuitive level.

Consider now the model

$$x_t = \phi x_{t-1} + \epsilon_t$$

where the independent chance elements ϵ_t are identically distributed with zero mean and variance σ^2, and ϕ is a parameter. This type of process is called an autoregressive process, since its form represents a regression of x_t on x_{t-1}. To illustrate the structure of the model, Table 7.1 shows the simulation of data from this model.

If we rewrite the expression for one time unit back,

$$x_{t-1} = \phi x_{t-2} + \epsilon_{t-1}$$

and substitute in the first expression, we get

$$x_t = \epsilon_t + \phi \epsilon_{t-1} + \phi^2 x_{t-2}$$

If we repeat this process over and over again, we eventually get

$$x_t = \epsilon_t + \phi \epsilon_{t-1} + \ldots + \phi^{t-2} \epsilon_2 + \phi^{t-1} \epsilon_1 + \phi^t x_0$$

where x_0 is the initial value of the sequence of x. Thus each x_t is a weighted sum of all past ϵ and the initial x_0. If t is large and $-1 < \phi < 1$, we may regard the sum as approximately infinite and

Table 7.1 Simulated autoregressive process
$\phi = 0.8, x_0 = 1$

t	ϕx_{t-1}	ϵ_t	$x_t = \phi x_{t-1} + \epsilon_t$
0			1.0
1	0.8	−0.465	0.335
2	0.268	−2.120	−1.852
3	−1.482	−2.748	−4.230
4	−0.384	0.308	−3.076
5	−2.461	−1.178	−3.639
6	−2.911	0.063	−2.848
7	−2.278	0.377	−1.901
8	−1.521	−1.412	−2.933
9	−2.346	1.322	−1.024
10	−0.820	0.204	−0.616

Table 7.2 A simulated moving average process
$\theta = 0.5$

t	ϵ_t	$\theta \epsilon_{t-1}$	$x_t = \epsilon_t - \theta \epsilon_{t-1}$
0	1.563		
1	1.085	0.7815	0.303
2	−0.345	0.5425	−0.887
3	−0.592	−0.1725	−0.419
4	0.399	−0.2960	0.798
5	0.568	0.1995	0.368
6	−0.377	0.2840	−0.661
7	−0.697	−0.1885	−0.508
8	−2.457	−0.3485	−2.108
9	0.410	−1.2285	1.638
10	−0.225	0.205	−0.430

write

$$x_t = \epsilon_t + \phi \epsilon_{t-1} + \phi^2 \epsilon_{t-2} + \ldots$$

As another example of a stochastic process consider the model

$$x_t = \epsilon_t - \theta \epsilon_{t-1}$$

The sequence of x values in this model can be imagined as being a weighted moving average (or rather moving sum), with weights $(1, -\theta)$, of a sequence of ϵ. This in fact is an example of what is called a moving average process. Table 7.2 shows an example of this type of process, again illustrated by showing a simulation.

It is seen that both autoregressive and moving average models can be expressed in terms of a weighted moving sum (either infinite or finite) of the sequence of ϵ.

It may help if we imagine these models to be formed by a physical process in which the ϵ_i values form the input to some mechanism or system and the x_i values the output. Suppose, for example, that when in a particular system there is only a single unit input at time 1 and no further inputs, then the system responds by giving outputs of 1 at time 1, ϕ at time 2, ϕ^2 at time 3, and so on, as in Table 7.3(a). The sequence 1, ϕ, ϕ^2, . . . is often referred to as the impulse response of the system. Consider the response of the system at time 3 to a random sequence ϵ_1, ϵ_2, ϵ_3. At time 3 the output will be composed of an output of ϵ_3 due to ϵ_3 arriving at time 3, $\phi\epsilon_2$ due to ϵ_2 arriving 2 and $\phi^2\epsilon_1$ due to ϵ_1 arriving at time 1, assuming here that the magnitude of the output is simply the impulse response times the size of the input. Assuming further that the actual ouput is the sum of the contributions of the responses to all past and present inputs, we arrive at a total output of

$$x_3 = \epsilon_3 + \phi\epsilon_2 + \phi^2\epsilon_1$$

A system for which we can carry out the above multiplications and additions is called a linear system and the models so defined are called linear models. Table 7.3(b) shows how the output sequence for this particular system is obtained from the input sequence for $t = 1$, 2, 3, ... It will be seen that for t tending to infinity we obtain for the output sequence the simple autoregressive model.

We have described a physical system in this example. There are many situations in business and economics where similar phenomena may be expected. The ϵ_t quantity is the total effect of the unpredictable factors in the situation at time t. The quantity ϵ_t will not only affect x_t but will also influence x_{t+1}, x_{t+2}, etc., though the influence will reduce as time passes, a reduction in our example governed by the parameter ϕ.

Consider now the problem in our example of forecasting x_4 and x_5 from the preceding data. The output at times 4 and 5 will depend on two types of input. Firstly, it will depend on the inputs at times 1, 2 and 3, whose effects we have already observed and, secondly, on the future inputs ϵ_4 and ϵ_5. We cannot attempt to forecast any particular numerical values for these inputs since they are independent of any existing information. The best we can do is to use their expected values, which as before we will take as zero. Thus we have to replace the known structure of Table 7.3(b) with a similar table, Table 7.3(c), which allows for our ignorance of future events. Thus our forecasts of x_4 and x_5 will be

$$\hat{x}_4 = 0 + \phi\epsilon_3 + \phi^2\epsilon_2 + \phi^3\epsilon_1$$
$$\hat{x}_5 = 0 + \phi^2\epsilon_3 + \phi^3\epsilon_2 + \phi^4\epsilon_1$$

In terms of the observations x_1, x_2, x_3 these will be

$$\hat{x}_4 = \phi x_3$$

Table 7.3 Inputs and outputs of a simple linear system

(a) Impulse response

Time	1	2	3	4	5	6
Input	1	0	0	0	0	0
Output	1	ϕ	ϕ^2	ϕ^3	ϕ^4	ϕ^5 $(-1 < \phi < 1)$

(b) Response to a sequence

Table elements give the response at time T due to input at time t.

Input times t	1	2	3	4	5	
Input	ϵ_1	ϵ_2	ϵ_3	ϵ_4	ϵ_5	Total output at time T x_T
Output times T						
1	ϵ_1					ϵ_1
2	$\phi\epsilon_1$	ϵ_2				$\epsilon_2 + \phi\epsilon_1$
3	$\phi^2\epsilon_1$	$\phi\epsilon_2$	ϵ_3			$\epsilon_3 + \phi\epsilon_2 + \phi^2\epsilon_1$
4	$\phi^3\epsilon_1$	$\phi^2\epsilon_2$	$\phi\epsilon_3$	ϵ_4		$\epsilon_4 + \phi\epsilon_3 + \phi^2\epsilon_2 + \phi^3\epsilon_1$
5	$\phi^4\epsilon_1$	$\phi^3\epsilon_2$	$\phi^2\epsilon_3$	$\phi\epsilon_4$	ϵ_5	$\epsilon_5 + \phi\epsilon_4 + \phi^2\epsilon_3 + \phi^3\epsilon_2 + \phi^4\epsilon_1$

(c) Response table at time $t = 3$

Input times t	1	2	3	4	5	
Input	ϵ_1	ϵ_2	ϵ_3			Total output at time T
Output times T						Actual
1 past	ϵ_1					ϵ_1
2	$\phi\epsilon_1$	ϵ_2				$\epsilon_2 + \phi\epsilon_1$
3 present	$\phi^2\epsilon_1$	$\phi\epsilon_2$	ϵ_3			$\epsilon_3 + \phi\epsilon_2 + \phi^2\epsilon_3$
						Forecasts Forecasts
4 future	$\phi^3\epsilon_1$	$\phi^2\epsilon_2$	$\phi\epsilon_3$	0		$\phi\epsilon_3 + \phi^2\epsilon_2 + \phi^3\epsilon_1$
5	$\phi^4\epsilon_1$	$\phi^3\epsilon_2$	$\phi^2\epsilon_3$	0	0	$\phi^2\epsilon_3 + \phi^3\epsilon_2 + \phi^4\epsilon_1$

and

$$\hat{x}_5 = \phi^2 x_3$$

Now we have a simple intuitive method of forecasting. Whether or not this is a good method is not obvious; in fact this gives forecasts with minimum mean square error for this model. To obtain the forecast error, we simply note that we have replaced the contributions in the outputs from ϵ_4 and ϵ_5 by zero and hence the errors are

$$e_4 = \epsilon_4$$

and

$$e_5 = \epsilon_5 + \phi\epsilon_4$$

as will be seen by comparing parts (b) and (c) of Table 7.3. The error variances, which here equal the mean square error as the expectations are zero, are thus

$$E(e_4^2) = \sigma^2$$

$$E(e_5^2) = (1 + \phi^2)\sigma^2$$

again using the independence of the ϵ values.

It should be clear from the above special case that for the autoregressive model the forecast with lead time h is

$$\hat{x}_{t+h} = \phi^h \epsilon_t + \phi^{h+1}\epsilon_{t-1} + \phi^{h+2}\epsilon_{t-2} + \ldots$$

which is

$$\hat{x}_{t+h} = \phi^h x_t$$

We may also derive this formula intuitively from the basic form of the model. The future observation x_{t+1} can be written as

$$x_{t+1} = \phi x_t + \epsilon_{t+1}$$

At time t, x_t is known but ϵ_{t+1} can at best be forecast as zero, so we have as a forecast

$$\hat{x}_{t+1} = \phi x_t$$

Similarly, the future value x_{t+2} can be written as

$$x_{t+2} = \phi x_{t+1} + \epsilon_{t+2}$$

At time t, x_{t+1} is not known but can be forecast by $\hat{x}_{t+1}$; ϵ_{t+2} again is forecast as zero. Thus

$$\hat{x}_{t+2} = \phi\hat{x}_{t+1}$$

and so

$$\hat{x}_{t+2} = \phi^2 x_t$$

Thus, in general,

$$\hat{x}_{t+h} = \phi^h x_t$$

For this forecast the error e_{t+h} is given by

$$e_{t+h} = \epsilon_{t+h} + \phi\epsilon_{t+h-1} + \ldots + \phi^{h-1}\epsilon_{t+1}$$

since the error consists of the part of the model of x_{t+h} involving the future unpredictable ϵ. The error has zero expectation so that the forecast is unbiased. The error variance is given by

$$\text{Var}(e_{t+h}) = \sigma^2 (1 + \phi^2 + \ldots + \phi^{2h-2})$$

and this can be used to calculate the necessary confidence limits for the forecast. If σ^2 is not known, we must consider how to estimate it. This can be done by noting that if we forecast only one step ahead, $h = 1$, then

$$e_{t+1} = \epsilon_{t+1}$$

If we go back through our past data and use the forecasting formula

$$\hat{x}_{t+1} = \phi x_t$$

then the set of forecast errors obtained will in fact be the actual random variable, ϵ. The variance of the e value is thus σ^2 and an estimate of σ^2 is provided by

$$\hat{\sigma}^2 = \frac{1}{t} \sum_{t=1}^{t} e_i^2$$

Table 7.4 shows the forecasting procedure for the autoregressive model discussed above. It is seen from this example that to start the process we need an initial value x_0. In the example of Table 7.4 we have forged data for which $x_0 = 1$ and $\phi = 0.8$, but in carrying out the forecasts we have assumed ϕ to be known to the forecaster but x_0 unknown and set at the value zero, just for the sake of illustration (in practice neither ϕ nor x_0 will be known). In the example the effect of the starting value dies away immediately and the forecast errors are equal to the ϵ values of the simulation. The real problem then is not of choosing x_0 but of estimating ϕ. The method of least squares can be

Table 7.4 Forecasting an autoregression process

Known model parameter $\phi = 0.8$
True initial condition $x_0 = 1.0$
Initial condition for forecasts $x_0 = 0$

Simulation of data			Forecast	Error
ϕx_{t-1}	ϵ_t	x_t	$\hat{x}_{t+1} = \phi x_t$	e_t
		1.0		
0.8	−0.465	0.335	0.0	0.335
0.268	−2.120	−1.852	0.268	−2.120
−1.482	−2.748	−4.230	−1.482	−2.748
−3.384	0.308	−3.076	−3.384	0.308
−2.461	−1.178	−3.639	−2.461	−1.178
−2.911	0.063	−2.848	−2.911	0.063
−2.278	0.377	−1.901	−2.278	0.377
−1.521	−1.412	−2.933	−1.521	−1.412

used for this purpose. The value of $\hat{\phi}$ is thus chosen to minimize the sum of residuals squared, i.e.

$$\sum_{i=2}^{t} \hat{\epsilon}_i^2 = \sum_{i=2}^{t} (x_i - \hat{\phi}x_{i-1})^2$$

The minimum is obtained when

$$\hat{\phi} = \sum_{i=2}^{t} x_i x_{i-1} \Big/ \sum_{i=2}^{t} x_{i-1}^2$$

The summation starts from $i = 2$ rather than $i = 1$, as x_0 is not an observation. As new data are obtained, both numerator and denominator can be modified by addition of the next available product and squared term ($x_{t+1}x_t$ and x_t^2), respectively. If we wish to regard our basic model as a local model, we may estimate ϕ using the method of discounted least squares. This has the effect of replacing the sums in the formula by weighted sums. The discounted estimate with exponential weights can be written as

$$\hat{\phi}_t = \sum_{r=0}^{t-2} a^r x_{t-r} x_{t-r-1} \Big/ \sum_{r=0}^{t-2} a^r x_{t-r-1}^2 = N_t / D_t$$

We may update both numerator and denominator by the usual type of recurrence relations:

$$\sum_{r=0}^{t-1} a^r x_{t+1-r}\, x_{t+1-r-1} = x_{t+1}x_t + a \sum_{r=0}^{t-2} a^r x_{t-r} x_{t-r-1}$$

i.e.

$$N_{t+1} = x_{t+1}x_t + aN_t \quad (N_1 = 0)$$

and

$$\sum_{r=0}^{t-1} a^r x_{t+1-r-1}^2 = x_t^2 + a \sum_{r=0}^{t-2} a^r x_{t-r-1}^2$$

so

$$D_{t+1} = x_t^2 + aD_t \quad (D_1 = 0)$$

Table 7.5 illustrates the use of this method on the same data as that in Table 7.4. It will be seen that the initial estimates of ϕ_t are wild, but soon settle down to vary about the true value used in the simulation. Similarly, the forecast errors soon become fairly close to the ϵ values used in the simulations of Table 7.1. We have taken the discounting factor a as 0.8 to illustrate the method, but clearly we are dealing with simulated data which we therefore know to come from a globally valid

100

Table 7.5 Forecasting the data of Table 7.4
ϕ estimated using discounted least squares, $a = 0.8$

x_t	x_{t-1}	N_t	D_t	$\hat{\phi}_t$	Forecast $\hat{\phi}_t x_t$	Error e_t
0.335						
−1.852	0.335	−0.620	0.112	—	—	—
−4.230	−1.852	7.338	3.520	—	—	—
−3.076	−4.230	18.882	20.709	0.91	—	—
−3.639	−3.076	26.299	26.029	1.01	−2.799	−0.840
−2.848	−3.639	31.403	34.065	0.92	−3.675	0.827
−1.901	−2.848	30.536	35.636	0.86	−2.620	0.719
−2.933	−1.901	30.005	30.866	0.97	−1.635	−1.298
−1.024	−2.933	27.007	33.295	0.81	−2.845	1.821
−0.616	−1.024	22.237	27.685	0.80	−0.829	0.213

model. Thus $a = 1$ would be the correct value to use; it would give the best estimator of ϕ and the best forecasts.

In this section we have concentrated on developing some intuitive ideas about forecasting stochastic processes. To progress further, we need a method for justifying these intuitive methods.

7.2 Forecasting and conditional expectation

In this section we discuss briefly how to find a forecasting formula, given observations $x_1, x_2, \ldots, x_t$, and the stochastic, or other, model underlying the data. Clearly, we would wish the forecasts obtained from such a formula to satisfy some criteria of quality. The one we will use is the mean square error. Thus we require an expression for the forecast $\tilde{x}_{t,h}$ of x_{t+h} for which

$$\text{MSE} = E\{(x_{t+h} - \tilde{x}_{t,h})^2\}$$

is as small as possible. The basic result that we will use is that the $\tilde{x}_{t,h}$ which minimizes the mean square error is the expected value of the future x_{t+h}, given that $x_1, x_2, \ldots, x_t$ are already known. This is referred to as the 'conditional expectation' and is denoted by

$$E(x_{t+h} \mid x_1, x_2, \ldots, x_t)$$

This result is derived in Appendix B. To see how to use this result, let us look again at the autoregressive model of the last section, namely

$$x_{t+1} = \phi x_t + \epsilon_{t+1} (t = 1, 0, \ldots)$$

We have written the model for x_{t+1} since the result above states that the forecast of x_{t+1} is its expectation, given $x_1, \ldots, x_t$. Formally

$$\tilde{x}_{t,1} = E(x_{t+1} \mid x_1, \ldots, x_t)$$

$$= E(\phi x_t + \epsilon_{t+1} \mid x_1, \ldots, x_t)$$

The conditional expectation of x_t, given the value of x_t, must be that given value, x_t. As the future ϵ values are independent of past events, the conditional expectation of ϵ_{t+1}, given $x_1, \ldots, x_t$, is just its ordinary expectation, which is zero. Thus

$$\tilde{x}_{t,1} = \phi x_t$$

which was the result argued intuitively in the last section.

7.3 Moving average processes

In section 7.1 we mentioned the model

$$x_t = \epsilon_t - \theta \epsilon_{t-1}$$

This process is an example of the moving average process. The general expression for such a process is

$$x_t = \epsilon_t - \theta_1 \epsilon_{t-1} - \theta_2 \epsilon_{t-2} - \ldots - \theta_q \epsilon_{t-q}$$

where $\theta_1, \theta_2, \ldots, \theta_q$ are constants. This will be referred to as a finite moving average process of order q, MA(q). The series might continue to infinity, in which case it would be termed as an infinite moving average process. We assume that either the expectation of x is zero naturally or if it was originally μ we subtract μ from each actual observation to obtain x. The forecasting of processes like this and other related processes to be discussed later in this chapter has been investigated in detail by G.E.P. Box and G.M. Jenkins (see Anderson, 1975; Box and Jenkins, 1970; Nelson, 1973); on the whole the notation of this chapter is that used by Box and Jenkins.

Let us start by considering a simple moving average process of order one:

$$x_t = \epsilon_t - \theta \epsilon_{t-1}$$

The minimum mean square error forecast of x_{t+1} is

$$\tilde{x}_t = E(x_{t+1} \mid x_t, x_{t-1}, \ldots)$$

As fixing the values of $x_t, x_{t-1}, \ldots$ is equivalent to fixing the values of $\epsilon_t, \epsilon_{t-1}, \ldots$, we can rewrite this as

$$\tilde{x}_t = E(x_{t+1} \mid \epsilon_t, \epsilon_{t-1}, \ldots)$$

$$= E(\epsilon_{t+1} - \theta \epsilon_t \mid \epsilon_t, \epsilon_{t-1}, \ldots)$$

$$= E(\epsilon_{t+1}) - \theta \epsilon_t$$

The future ϵ_{t+1} has zero expectation and $E(\epsilon_t | \epsilon_t)$ is the expectation of a constant, ϵ_t. Thus $\tilde{x}_t = -\theta\epsilon_t$, and we obtain a formula for the forecast in terms of ϵ_t. To obtain a more practical forecast, we first examine the prediction error

$$\begin{aligned} e_{t+1} &= x_{t+1} - \tilde{x}_t \\ &= (\epsilon_{t+1} - \theta\epsilon_t) - (-\theta\epsilon_t) \\ &= \epsilon_{t+1} \end{aligned}$$

We again have a situation in which the one-step-ahead forecast errors, which we can obviously determine directly, are identical to the ϵ values in the model. The forecasting formula can thus be rewritten as

$$\tilde{x}_t = -\theta e_t$$

which is also expressible in the recurrence form

$$\tilde{x}_t = -\theta(x_t - \tilde{x}_{t-1})$$

If we seek to express $\tilde{x}_t$ directly in terms of the data, we may repeatedly substitute for $\tilde{x}_{t-1}$, $\tilde{x}_{t-2}$, ... in the last expression, giving

$$\tilde{x}_t = -\theta x_t - \theta^2 x_{t-1} - \theta^3 x_{t-2} - \ldots$$

It follows from this last expression that to make sense of our forecast we must require that θ lies in the interval $(-1, 1)$. Finally, as $e_t = \epsilon_t$ the minimum mean square error is in fact σ^2. Table 7.6 shows a set of data and forecasts for this model, in which e_0 has been taken as zero to provide a starting value.

Suppose now that we wish to forecast two time units into the future.

Table 7.6 Forecasting a moving average
process

$$\tilde{x}_t = -\theta e_t, \theta = 0.5, e_0 = 0$$

t	x_t	$\tilde{x}_{t-1}$	e_t	$\tilde{x}_t$
0				
1	0.303	0	0.303	−0.1515
2	−0.887	−0.151	−0.736	0.368
3	−0.419	0.368	−0.787	0.393
4	0.798	0.393	0.405	−0.202
5	0.368	−0.202	0.570	−0.285
6	−0.661	−0.285	−0.376	0.188
7	−0.508	0.188	−0.696	0.348
8	−2.108	0.348	−2.456	1.228
9	1.638	1.228	0.410	−0.205
10	−0.430	−0.205	−0.225	0.112

From our model

$$x_{t+2} = \epsilon_{t+2} - \theta\epsilon_{t+1}$$

on the right-hand side of which we have only future values of ϵ. It follows that

$$E(x_{t+2} \mid \epsilon_t, \epsilon_{t-1}\ldots) = E(\epsilon_{t+2}) - \theta E(\epsilon_{t+1}) = 0$$

Thus we are unable to obtain a better predictor of x_{t+2} than its mean value of zero, and the same obviously applies to any x more than one time unit into the future.

The generalization of the above example to obtain forecasting formulae for the moving average process of order q is quite straightforward. The forecast is given by the expectation of x_{t+1} conditional on x_t, x_{t-1}, ..., which is equivalent to conditioning on $\epsilon_t, \epsilon_{t-1}, \ldots$ Hence

$$\tilde{x}_t = E(\epsilon_{t+1} - \theta_1\epsilon_t - \theta_2\epsilon_{t-1} - \ldots - \theta_q\epsilon_{t+1-q} \mid \epsilon_t, \epsilon_{t-1}, \ldots)$$

so

$$\tilde{x}_t = -\theta_1\epsilon_t - \theta_2\epsilon_{t-1} - \ldots - \theta_q\epsilon_{t+1-q}$$

The forecast error is

$$e_{t+1} = x_{t+1} - \tilde{x}_t$$

which, on expressing both x_{t+1} and $\tilde{x}_t$ in terms of ϵ values, gives

$$e_{t+1} = \epsilon_{t+1}$$

So once again the one-step-ahead forecast errors correspond to the ϵ values and the minimum mean square error is σ^2. Using this fact to obtain a more practical form of forecast, we have

$$\tilde{x}_t = -\theta_1 e_t - \theta_2 e_{t-1} - \ldots - \theta_q e_{t+1-q}$$

where the e values are past forecast errors.

The standard deviation of the prediction error is σ and so confidence limits for x_{t+1}, given our forecast, can be obtained directly. For example, the 95 per cent confidence limits would be $\tilde{x}_t \pm 1.96\,\sigma$, assuming a normal distribution. Where σ is not known, it is reasonable to estimate it directly using the sample standard deviation of the forecast errors.

If we now seek to predict two steps ahead, the argument used before gives

$$\tilde{x}_{t,2} = -\theta_2\epsilon_t - \theta_3\epsilon_{t-1} - \ldots - \theta_q\epsilon_{t+2-q}$$

or

$$\tilde{x}_{t,2} = -\theta_2 e_t - \theta_3 e_{t-1} - \ldots - \theta_q e_{t+2-q}$$

where the e values are still the one-step-ahead forecast errors. Comparing the forms of $\tilde{x}_t$ and $\tilde{x}_{t,2}$, it is seen that if we are forecasting both x_{t+1} and x_{t+2} at each stage we can simply update our forecast $\tilde{x}_t$ using

$$\tilde{x}_{t+1,1} = \theta_1 e_{t+1} + \tilde{x}_{t,2}$$

We can continue generating forecasts until we seek to forecast x_{t+q+1} for which the basic model gives

$$x_{t+q+1} = \epsilon_{t+q+1} - \theta_1 \epsilon_{t+q} - \ldots - \theta_q \epsilon_{t+1}$$

As there are only future ϵ values on the right-hand side, the conditional expectation of x_{t+q+1} is simply zero. Thus we have

$$\tilde{x}_{t,q+k} = 0 \quad (k \geqslant 1)$$

The forecast error for forecasting $h(h \leqslant q)$ steps into the future is

$$e_{t+h} = x_{t+h} - \tilde{x}_{t,h}$$
$$= (\epsilon_{t+h} - \theta_1 \epsilon_{t+k+1} - \ldots - \theta_{h-1}\epsilon_{t+1} - \theta_h \epsilon_t - \ldots - \theta_q \epsilon_{t+h-q})$$
$$- (\qquad\qquad\qquad\qquad\qquad - \theta_h \epsilon_t - \ldots - \theta_q \epsilon_{t+h-q})$$
$$= \epsilon_{t+h} - \theta_1 \epsilon_{t+h-1} - \ldots - \theta_{h-1} \epsilon_{t+1}$$

i.e. it equals the 'future' part of the model for x_{t+h}. This has zero expectation and variance

$$\sigma^2_{t,h} = \sigma^2 (1 + \theta_1^2 + \ldots + \theta_{h-1}^2)$$

If we wish to give 'confidence' limits for x_{t+h}, given our forecast, we have, for example, that the 95 per cent confidence interval, or prediction interval, is

$$\tilde{x}_{t,h} \pm 1.96\, \sigma_{t,h}$$

assuming normality in the ϵ values.

The above discussion was concerned with models only. To make this method practical, we need to be able to estimate the parameters $\theta_1, \ldots, \theta_q$. The theory of this estimation is beyond the scope of this book, but is dealt with in detail in the book by Box and Jenkins, using the method of least squares. The use of discounted least squares is also possible. The basic problem with both methods is that the model does not enable the residual sum of squares, $\Sigma \hat{e}^2$ (discounted or otherwise), to be written as a simple function of the parameters. Thus we have nothing corresponding to the normal equations that can be simply solved. The methods needed are thus fairly complicated and involve repeated approximations to the answer. The methods do not enable the parameters to be re-estimated, as new data is obtained, in a simple recursive fashion.

By repeated substitution using

$$e_{t+1} = \epsilon_{t+1} = x_{t+1} - \tilde{x}_t$$

in both models and forecasts for moving average processes, we find that both can be expressed as an infinite series of past observations, whose coefficients depend on $\theta_1, \ldots, \theta_q$. Experience suggests that with many sets of data the first few terms of these series can be used to provide good forecasts. We can thus use an autoregressive form of model as an approximation to a moving average model. This avoids the problems of estimation referred to above, since, as we have already illustrated in section 7.1, the autoregressive parameters can be directly estimated.

7.4 Autoregressive processes

The general pth order autoregressive process is defined by

$$x_t - \phi_1 x_{t-1} - \phi_2 x_{t-2} - \ldots - \phi_p x_{t-p} = \epsilon_t$$

where $\phi_1, \phi_2, \ldots, \phi_p$ are constants and the model is denoted AR(p). To investigate the forecasting of such a process we start with the simplest case, the first-order autoregressive process, which we have already studied in some detail in section 7.1. The model is

$$x_t = \phi_1 x_{t-1} + \epsilon_t$$

We have already shown in section 7.2 that the minimum mean square error forecast of x_{t+1} is

$$\tilde{x}_t = \phi_1 x_t$$

The forecast error is

$$\begin{aligned} e_{t+1} &= x_{t+1} - \tilde{x}_t \\ &= (\phi_1 x_t + \epsilon_{t+1}) - \phi_1 x_t \\ &= \epsilon_{t+1} \end{aligned}$$

So the forecast error is the future ϵ_{t+1}, and the mean square error is thus σ^2.

To forecast two steps into the future, we obtained in section 7.1

$$\tilde{x}_{t,2} = \phi_1^2 x_t$$

As a somewhat different way of getting this result we could write

$$\begin{aligned} \tilde{x}_{t,2} &= E(\phi_1 x_{t+1} + \epsilon_{t+2} \mid x_t, x_{t-1}, \ldots) \\ &= \phi_1 E(x_{t+1} \mid x_t, x_{t-1}, \ldots) \end{aligned}$$

which is

$$\tilde{x}_{t,2} = \phi_1 \tilde{x}_t$$

and hence

$$\tilde{x}_{t,2} = \phi_1^2 x_t$$

The prediction interval may be found directly by writing

$$e_{t+2} = (\phi_1^2 x_t + \phi_1 \epsilon_{t+1} + \epsilon_{t+2}) - \phi_1^2 x_t$$

$$= \phi_1 \epsilon_{t+1} + \epsilon_{t+2}$$

So

$$\mathrm{Var}(e_{t+2}) = \sigma_{t,1}^2 = (\phi_1^2 + 1)\sigma^2$$

Hence, assuming normality, the 95 per cent prediction interval for x_{t+2} is

$$\phi_1^2 x_t \pm 1.96(\phi_1^2 + 1)\sigma^2$$

For the general pth-order autoregressive processes we have

$$x_{t+1} = \phi_1 x_t + \phi_2 x_{t-1} + \ldots + \phi_p x_{t+1-p} + \epsilon_{t+1}$$

and so, as

$$\tilde{x}_{t,1} = E(x_{t+1} \mid x_t, \ldots, x_1)$$

we have

$$\tilde{x}_{t,1} = \phi_1 x_t + \phi_2 x_{t-1} + \ldots + \phi_p x_{t+1-p}$$

as the one-step-ahead forecast and

$$e_{t+1} = x_{t+1} - \tilde{x}_{t,1} = \epsilon_{t+1}$$

as the one-step-ahead forecast error. Thus the model provides the forecast in a very direct fashion and once again the ϵ values are equal to the one-step-ahead prediction errors.

For forecasting two steps ahead,

$$\tilde{x}_{t,2} = E(x_{t+2} \mid x_t, \ldots, x_1)$$

$$= E(\phi_1 x_{t+1} + \phi_2 x_t + \phi_3 x_{t-1} + \ldots + \phi_p x_{t+2-p} + \epsilon_{t+2} \mid x_t, \ldots, x_1)$$

$$= \phi_1 E(x_{t+1} \mid x_t, \ldots, x_1) + \phi_2 x_t + \ldots + \phi_p x_{t+2-p}$$

$$\tilde{x}_{t,2} = \phi_1 \tilde{x}_{t,1} + \phi_2 x_t + \ldots + \phi_p x_{t+2-p}$$

By a similar argument

$$\tilde{x}_{t,3} = \phi_1 \tilde{x}_{t,2} + \phi_2 \tilde{x}_{t,1} + \phi_3 x_t + \ldots + \phi_p x_{t+3-p}$$

Thus in predicting a future value x_{t+h} we first write down the model for x_{t+h} and delete the term ϵ_{t+h}. When an x in the model refers to a future value, we replace it by its predicted value, $\tilde{x}$; where it refers to a past value we simply use the observed value. To find the prediction errors of the above, we subtract $\tilde{x}_{t,2}$ from x_{t+2}, giving

$$e_{t+2} = \epsilon_{t+2} + \phi_1(x_{t+1} - \tilde{x}_{t,1}) = \epsilon_{t+2} + \phi_1 e_{t,1}$$

and subtract $\tilde{x}_{t,3}$ from x_{t+3}, giving

$$e_{t,3} = \epsilon_{t+3} + \phi_1(x_{t+2} - \tilde{x}_{t,2}) + \phi_2(x_{t+1} - \tilde{x}_{t,1})$$

$$= \epsilon_{t+3} + \phi_1 e_{t,2} + \phi_2 e_{t,1}$$

and so

$$e_{t,3} = \epsilon_{t+3} + \phi_1 \epsilon_{t+2} + \phi_1^2 \epsilon_{t+1} + \phi_2 \epsilon_{t+1}$$

We can use relationships such as this to find prediction intervals for different lead times. Thus

$$\sigma_{t,2}^2 = \sigma^2 + \mathrm{Var}(e_{t,1}) = \sigma^2(1 + \phi_1^2)$$

and

$$\sigma_{t,3}^2 = \sigma^2\{1 + \phi_1^2 + (\phi_1^2 + \phi_2)^2\}$$

The above discussion assumes that we know the parameters $\phi_1 \ldots,$ ϕ_p. To obtain practical forecasts, we must clearly obtain estimates of these from the data. As the model is of linear regression form, we may use the methods of least squares or discounted least squares described in the previous chapter to obtain appropriate normal equations. A simple illustration of this was given for the model AR(1) in section 7.1 and illustrated in Table 7.5.

7.5 Autoregressive — moving average models

Box and Jenkins (1970) have considered a generalization of the two previous models. The reader is referred to this reference for details. The aim here is to illustrate the basic ideas of this more general model by considering some of the simpler cases.

The most natural generalization of the two previous models is to combine them. Thus if we take the first-order moving average process

$$x_t = \epsilon_t - \theta\epsilon_{t-1}$$

and the first-order autoregressive process

$$x_t - \phi x_{t-1} = \epsilon_t$$

and combine their structures, we obtain

$$x_t - \phi x_{t-1} = \epsilon_t - \theta\epsilon_{t-1}$$

This would be referred to as a 'mixed moving average — autoregressive process' of order (1, 1).

To forecast x_{t+1} we note that

$$x_{t+1} = \phi x_t + \epsilon_{t+1} - \theta \epsilon_t$$

So taking the conditional expectation at time t gives

$$\tilde{x}_t = \phi x_t - \theta \epsilon_t$$

The forecast error is

$$e_{t+1} = x_{t+1} - \tilde{x}_t = \epsilon_{t+1}$$

so once again the one-step-ahead forecast errors are equal to the corresponding ϵ values. Thus for practical use we may write

$$\tilde{x}_t = \phi x_t - \theta e_t$$

or in terms of the previous forecast

$$\tilde{x}_t = (\phi - \theta)x_t + \theta \tilde{x}_{t-1}$$

If $\phi = 1$ and $0 < \theta < 1$, this is identical to the forecasting formula of exponential smoothing; thus exponential smoothing is the minimum mean square error forecasting method for the model

$$x_t = x_{t-1} + \epsilon_t - \theta \epsilon_{t-1}$$

Here each value of x_t equals the previous value x_{t-1} plus a random term from a simple moving average stochastic process.

Consider now a prediction for a lead time of two in the original model

$$\tilde{x}_{t,2} = E(\phi x_{t+1} + \epsilon_{t+2} - \theta \epsilon_{t+1} \mid x_t, \ldots,)$$
$$= \phi E(x_{t+1} \mid x_t \ldots)$$
$$= \phi \tilde{x}_{t,1}$$

This same calculation obviously applies to any lead time, k say. Thus

$$\tilde{x}_{t,k} = \phi \tilde{x}_{t,k-1} \quad (k \geqslant 2)$$

so

$$\tilde{x}_{t,k} = \phi^{k-1} \tilde{x}_{t,1}$$

Considering now the problem of updating the forecasts, substitute for $\tilde{x}_{t,1}$ in this last expression; then

$$\tilde{x}_{t,k} = \phi^{k-1}(\phi - \theta)x_t + \theta \phi^{k-1} \tilde{x}_{t-1,1}$$

so

$$\tilde{x}_{t,k} = \theta \tilde{x}_{t-1,k} + \phi^{k-1}(\phi - \theta)x_t$$

Thus there are very simple recurrence relations for updating forecasts.

As with autoregressive and moving average models, we may generalize the above model to form

$$x_t - \phi_1 x_{t-1} - \phi_2 x_{t-2} - \ldots - \phi_p x_{t-p} = \epsilon_t - \theta_1 \epsilon_{t-1} - \theta_2 \epsilon_{t-2}$$
$$- \ldots - \theta_q \epsilon_{t-q}$$

This is referred to as an autoregressive moving average model of order (p,q), denoted by ARMA(p,q). The general procedure for forecasting with such a model follows exactly the lines indicated above.

At first sight a mixed model may appear to be a rather complicated statistician's toy without much intuitive basis. It can be shown, however, that we can expect such models to be appropriate in a range of practical situations. These include

(a) situations where the series of interest are an aggregate of a number of other series following simpler stochastic models;
(b) situations where a simpler model is observed subject to error;
(c) situations where an original simpler series is aggregated over time, e.g. where weekly data are aggregated to give monthly or yearly totals;
(d) situations where the series of interest are subject to feedback effects through their interactions with another variable.

Let now summarize the rules for writing down forecasting formulae for moving average, autoregressive and mixed models. To forecast x_{t+k}, write out the expression for x_{t+k} as given by the model.

This will be a linear function like

$$x_{t+k} = \text{terms in } x_{t+k-1}, \ldots, x_{t+1}, x_t, x_{t-1}, \ldots, \epsilon_{t+k}, \ldots, \epsilon_{t+1}, \epsilon_t,$$
$$\epsilon_{t-1}, \ldots$$

then $\tilde{x}_{t,k}$ is obtained directly from the right-hand side of this by

(a) replacing all future ϵ values by 0, i.e. $\epsilon_{t+1} = \ldots = \epsilon_{t+k} = 0$;
(b) replacing all past ϵ values by the corresponding forecast error of lead one, i.e. ϵ_T by $e_T = x_T - \tilde{x}_{T-1,1}$,
(c) replacing all future x values by their forecasts, i.e x_{t+h} by $\tilde{x}_{t,h}$;
(d) leaving all past x values at their observed values.

Example Consider prediction for the model

$$x_t - 0.5 x_{t-1} + 0.3 x_{t-2} = \epsilon_t - 0.8 \epsilon_{t-1}$$

(a) Thus

$$x_{t+1} = 0.5 x_t - 0.3 x_{t-1} + \epsilon_{t+1} - 0.8 \epsilon_t$$

so

$$\tilde{x}_{t,1} = 0.5 x_t - 0.3 x_{t-1} - 0.8 e_t$$

(b) $x_{t+2} = 0.5\, x_{t+1} - 0.3\, x_t + \epsilon_{t+2} - 0.8\, \epsilon_{t+1}$

so

$$\tilde{x}_{t,2} = 0.5\, \tilde{x}_{t,1} - 0.3\, x_t$$

(c) $x_{t+h} = 0.5\, x_{t+h-1} - 0.3\, x_{t+h-2} + \epsilon_{t+k} - 0.8\, \epsilon_{t+h-1}\; (k > 3)$

so

$$\tilde{x}_{t,h} = 0.5\, \tilde{x}_{t,h-1} - 0.3\, \tilde{x}_{t,h-2}$$

Using the extreme simplicity of these relations, we may also derive expressions for the forecast error variance for different lead times. Thus subtracting the forecast formulae from the models gives

(a) $e_{t+1} = \epsilon_{t+1}$, for a lead time of one,

so

$$\sigma_{t,1}^2 = \sigma^2$$

(b) For forecasts $\tilde{x}_{t,2}$, with lead two, the error is

$$
\begin{aligned}
e_{t+2} &= 0.5\, e_{t,1} + \epsilon_{t+2} - 0.8\, \epsilon_{t+1}\\
&= -0.3\, \epsilon_{t+1} + \epsilon_t
\end{aligned}
$$

so

$$\sigma_{t,2}^2 = 1.09\, \sigma^2$$

(c) For a lead time of k

$$e_{t+h} = 0.5\, e_{t+h-1} - 0.3\, e_{t+h-2} + \epsilon_{t+h} - 0.8\, \epsilon_{t+h-1}$$

Here e_{t+h-1} is not independent of e_{t+h-2} and so we cannot simply relate $\sigma_{t,h}^2$ to $\sigma_{t,k-1}^2$ and $\sigma_{t,h-2}^2$. By repeated use we may relate e_{t+3}, e_{t+4}, etc., to the ϵ values and hence find $\sigma_{t,3}^2, \sigma_{t,4}^2$, etc.

As usual we have assumed known parameter values. These models pose difficult problems of estimation, though computer programs to carry out the estimation are becoming increasingly common. Unfortunately, few of these programs update the estimates of the parameters as new data become available. They tend to keep to the parameters estimated from the original set of data, but include some form of quality control of the forecasts to indicate when the model may have changed. Experience, and intuition, suggest that updating the parameter estimates as new data arise lead to better forecasts. It should also be noted that where discounted least squares is used forecasting parameters of 0.95 and above commonly produce the best forecasts. We have also assumed in the discussions of this chapter that we have identified an appropriate model for the data in hand. The identification of stochastic models requires considerable data and experience and the details are beyond the scope of this book. Section 12.2 and appendix A give some indication of the approach but the reader is referred to the books by Box and Jenkins (1970) and Anderson (1975).

7.6 Models involving stochastic and deterministic elements

In the previous sections we have been examining stochastic models which contained no deterministic part. This is not very realistic, as at the least the data will have some non-zero mean which may or may not be globally constant. We therefore need to introduce methods for dealing with situations that show both stochastic structure and deterministic elements such as means and trends. There are two main approaches that have been introduced for such situations:

7.6.1 methods based on the elimination of deterministic elements from the data before the use of stochastic models;
7.6.2 methods based on the formulation of models that, though stochastic and global in nature, correspond to local deterministic structures.

Let us illustrate these two approaches in turn.

7.6.1 Elimination of deterministic elements

7.6.1 Suppose we look at a simple linear trend model

$$x_t = \alpha + \beta t + \epsilon_t$$

The trend part of this model can be eliminated by a process of differencing. That is to say, by using the differences between successive values, an operation that we denote by ∇. Thus differencing the data once gives

$$\nabla x_t = x_t - x_{t-1}$$
$$= \beta + \epsilon_t - \epsilon_{t-1}, \text{ in terms of the model}$$

Repeating this, and using $\nabla^2 x_t$ to denote the repetitions, we get

$$\nabla^2 x_t = \nabla x_t - \nabla x_{t-1}$$
$$= \epsilon_t - 2\epsilon_{t-1} + \epsilon_{t-2}, \text{ from the model}$$

Thus if we define a new variable w_t by

$$w_t = \nabla^2 x_t$$

the model becomes

$$w_t = \epsilon_t - 2\epsilon_{t-1} + \epsilon_{t-2}$$

The model for w_t is therefore a moving average model of order two. Thus the structure of a linear trend model can be incorporated into a moving average model by the use of differenced data $\nabla^2 x_t$ in place of the actual data. This suggests that very general models of a non-stationary nature can be constructed from moving average models, autoregressive models and the mixed moving average—autoregressive model by replacing the x_t by differences of x_t, say the dth difference $\nabla^d x_t$. Such models are termed 'integrated' models of order (p, d, q), p referring to the order of the autoregressive part of the model, d to the degree of differencing and q to the order of the moving average part of the model.

Consider, by way of further illustration, the integrated moving average model of order $(., 1, 1)$, i.e.

$$\nabla x_t = \epsilon_t - \theta \epsilon_{t-1}$$

Writing this as

$$x_t = x_{t-1} + \epsilon_t - \theta \epsilon_{t-1}$$

gives a form identical to the particular model discussed in section 7.5. As any difference operator $\nabla^d x_t$ can be expanded in terms of x_t, $x_{t-1}, \ldots, x_{t-d}$, it is clear that any integrated model can be expressed in autoregressive terms.

As the differencing uses only adjacent values of x, it is unaffected by slow variation in the underlying mean. Similarly, a model using $\nabla^2 x_t$ will be almost unaffected by slow variation in both mean and trend. The practical procedure here is to difference the data until a stable series of zero mean and small variance is obtained. If the series is differenced too many times, the variance will increase and the process of identifying and fitting the model for the differenced data will get very difficult.

The use of differencing may in practice suffer from two disadvantages. Firstly, to obtain the forecast of a future x_{t+1}, for example, we need to forecast ∇x_{t+1} by $\hat{\nabla} x_{t+1}$, and then 'undifference' to be able to write

$$\hat{x}_{t+1} = \hat{\nabla} x_{t+1} + x_t$$

Thus x_t contains the measure of underlying mean, since we are not explicitly estimating μ_t, the mean at time t. This may increase the root mean square error above that of a method that explicitly estimates the local value μ_t. As always, this can only be tested by trial and error. The other disadvantage is that often knowledge of the local mean and trend is an important side product of the forecasting exercise. This knowledge is not provided when differences are used. This disadvantage is overcome in the second approach to this problem.

7.6.2 Stochastic models with deterministic elements

The second approach seeks to formulate the model in a form that explicitly includes the deterministic parameters. Three particular models are of value here:

(i) If $E(x_t) = \mu$, for all time, then all the previous stochastic models can be straightforwardly reformulated with $(x_t - \mu)$, $(x_{t-1} - \mu)$, etc., replacing x_t, x_{t-1} .... Estimation of μ is not very difficult, given a good quantity of data.

(ii) If a constant trend occurs with an autoregressive model, it can be allowed for by the addition of a constant term in the regression form. Thus, for example,

$$x_t = \phi_1 x_{t-1} + \epsilon_t$$

would include a slope β if it was modified to

$$x_t = \beta + \phi_1 x_{t-1} + \epsilon_t$$

Again the estimation and forecasting procedures would be virtually unchanged by this additional term.

(iii) The third type of model gives a formulation that is stochastic and global in nature but which allows for local variation in mean and slope. We start with the 'local' constant mean model

$$x_t = \mu_t + \epsilon_t$$

where

$$E(\epsilon_t) = 0 \quad \text{and} \quad \text{Var}(\epsilon_t) = \sigma_\epsilon^2$$

We dealt with this model in section 4.2 by assuming that μ_t wandered in an unknown, undefined, fashion. We now formulate a model that describes how μ_t changes. If there is no trend, we use

$$\mu_t = \mu_{t-1} + \gamma_t$$

where

$$E(\gamma_t) = 0 \quad \text{and} \quad \text{Var}(\gamma_t) = \sigma_\gamma^2$$

This represents a mean that is constantly modified by the random variations, γ_t. It can be shown that, for large t, the forecast leading to the minimum mean square error for this model is just exponential smoothing. The forecasting parameter a of the exponential smoothing is, however, no longer an empirically introduced parameter; it is related directly to the model parameters σ_ϵ^2 and σ_γ^2 by

$$a = [(1 + 2R) - \surd(1 + 4R)]/2R \quad (R = \sigma_\epsilon^2/\epsilon_\gamma^2)$$

Thus, for example, if σ_γ^2 approaches zero, then a approaches 1, which corresponds to the local model approaching the global case of a fixed μ.

The values of σ_ϵ^2 and σ_γ^2 can be estimated from the correlation structure of the original data (see Grunwald, 1965).

The above model can be extended to allow for trend by using the ideas developed for Holt's method and illustrated in Figure 5.5. Thus the mean model is rewritten as

$$\mu_t = \mu_{t-1} + \beta_t + \gamma_t$$

and a wandering slope is modelled by

$$\beta_t = \beta_{t-1} + \delta_t$$

where

$$E(\delta_t) = 0 \quad \text{and} \quad \text{Var}(\delta_t) = \sigma_\delta^2$$

It can be shown (Harrison, 1967) that Holt's method provides the optimum (MSE) forecast for large t, for data from this model. Again the forecast parameters are related directly to the model parameters, σ_ϵ^2, σ_γ^2 and σ_δ^2. In a paper by Harrison and Stevens (1971) methods for estimating and updating the parameters are described. These allow also for the error variance and the best forecasting parameters to be updated as the process develops. As these methods depend on the use of a formula and approach developed by Rev. T. Bayes, over a century ago, they are often referred to as Bayesian forecasting methods. The technical details are beyond the scope of this book, but a further facet of the approach is discussed in sections 10.3 and 14.4.

It can be shown that the above models correspond to particular moving average models for ∇x and $\nabla^2 x$, respectively. The difference lies in the formulation of the models. For the above formulation μ and β are explicitly estimated by the forecaster, which is not the case in the differenced forms.

References

Anderson, O. D. (1975). *Time Series Analysis and Forecasting, the Box—Jenkins Approach*. Butterworths, London.

Box, G. E. P., and Jenkins, G. M. (1970). *Time Series Analysis, Forecasting and Control*. Holden-Day, Inc., San Francisco.

Grunwald, H. (1965). The correlation theory for stationary stochastic processes applied to exponential smoothing. *Statistica Nierlandica*, 19, 129—138.

Harrison, P. J. (1967). Exponential smoothing and short-term sales forecasting. *Manag. Sci.*, 13, No. 11, 821—842.

Harrison, P. J., and Stevens, C. F. (1971). A Bayesian Approach to short-term forecasting. *Op. Res. Quarterly*, 22, No. 4, 341—362.

Nelson, C. R. (1973). *Applied Time Series Analysis for Managerial Forecasting*. Holden-Day, Inc., San Francisco.

Chapter 8

Seasonal models

8.1 Introductory concepts

8.1.1 Seasonal structures

In this chapter we study forecasting using seasonal models, i.e. models which show some form of variation over each year which repeats, at least approximately, year by year. The year may be 52 weeks, 12 calendar months or 13 four-weekly periods. We may also apply the methods to variation within a working week which repeats weekly or any other variation which has a known fixed period. The basic cause of seasonal variations is usually either the seasonal weather, as in, say, airline bookings or sales of sun tan lotion or cough mixture, or the fixed seasonal events such as Christmas, motor shows and sales. However, the cause may be far removed from the series being studied.

Figure 8.1 shows some examples of various types of periodic variation. A major problem facing the forecaster of seasonally varying data is that of choosing the right type of model, since there are many ways in which seasonal structures occur. For the present it is sufficient to observe from Figure 8.1 that a seasonal structure may be constant from year to year or it may show some deviation from one year to the next. This deviation may be in the underlying mean level of the variation, as in Figure 8.1(b), or in the positioning of the features of the variation within the year, as in (c), or in the magnitudes of the variation, as in (d). In practice, of course, the deviation may be of any or all of these forms. To make the models realistic we must also assume that random variations also occur in our data.

In this chapter we will survey some of the most frequently used methods of analysing and forecasting data with seasonal features. Rather than look at each of these methods separately, we will seek to emphasize the common structure that most of these methods have and show the various techniques in these methods as particular approaches

116

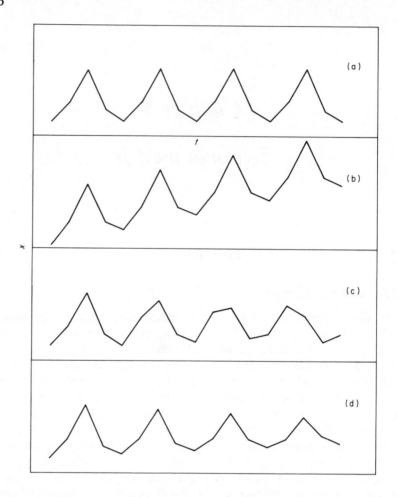

Figure 8.1   Types of seasonal structure (a) (b) (c) (d)

to solving certain common problems. The one common element in all the methods discussed in this chapter is their basic assumption as to what constitutes a general seasonal model. They differ in the way the model is expressed, but all see in it at least three of four components. These are explained briefly below.

(a) The trend component T

This is the long-term upward and downward movement of the series due to slow changes in factors that influence the mean of the series. Depending on the nature of the series, these factors might be, for example, growth in the firm, economic change, population change and change in the quality of materials.

(b) The seasonal component S

This is some measure of the characteristic behaviour of the series during each season in the period, e.g. each month in the year. The seasonal may be assumed to repeat itself in each period, though often we allow for slow changes in its structure.

(c) The cyclical component C

This has something of the character of the seasonal component in that it is expected to show upward and downward swings, but the term 'cycle' is kept to describe movements which are slower and much less predictable than the seasonal component. The length of the cycle and its peak magnitude are all uncertain. The length of many cycles average at about 3 to 4 years, though some are longer than 15 years. The nature and predictability of cycles has been under detailed discussion for many years. The reader is referred to the book by Dauten (1961) and Davis (1941) for a thorough treatment of this component. When considering short-term forecasting, it is convenient to consider the trend and the cycle together. The reasoning for this is simply that it is impossible to pick out useful information on cycles without data from several cycles, which is usually at least 10 years. Any cyclic feature in data obtained over shorter periods thus gets interpreted as part of the trend movement. Conversely, the trend itself can be regarded as part of a cycle of very long duration. In short-term forecasting the cycle is usually included in with the trend and in this chapter we will assume that the term T includes C in the short term.

(d) The irregular component I

This is, in effect, defined as the remainder when T, S and C have been accounted for. It will usually be regarded as being a random sequence. It may, however, be that some of its peaks and troughs can be explained in terms of the origin of the data. A trough corresponding to low sales might be due to some breakdown in delivery and the following peak due to the making up of the backlog. A valuable part of the analysis of data using this model is the increase in understanding of the situation which comes from this type of investigation of possible cause and effect.

We now have a set of possible components for a seasonal model — T, S, I. There are several ways in which these components could be put together to form a model for a set of seasonal data:

$T + S + I$

$T \times S \times I$

$T \times S + I$

The first model above is called the additive model. If x_{ij} is the observation for season j in period i, then the model is formally

$$x_{ij} = T_{ij} + S_j + I_{ij}$$

The second model above is called the pure multiplicative model:

$$X_{ij} = T_{ij} \, S_j \, I_{ij}$$

The third example is also a form of multiplicative model, but it has an additive irregular component. It is usual to say that a model is of an additive or multiplicative type according to the way the seasonal component acts on the trend/cycle terms.

What evidence there is available seems to suggest that the multiplicative form of model occurs most frequently in practice. Most commercially available programmes on seasonal forecasting take this model as their basis. It is, however, necessary to check on the form of model most appropriate for the data on hand.

If we take logarithms of the pure multiplicative model, we get

$$\log x_{ij} = \log T_{ij} + \log S_j + \log I_{ij}$$

$$= T_{ij}^1 + S_j^1 + I_{ij}^1$$

So the logarithms of the original data can be modelled by a pure additive model. We can thus carry out an analysis of the original data using the multiplicative model or of the transformed data using the additive model. The distribution of I_{ij} often tends to be highly skew. A property of the log transformation is that, if this is the case, $\log I_{ij}$ tends to be much less skew and may even be approximately normally distributed. In cases where a large irregular component occurs this symmetry of distribution has the advantage in forecasting that symmetric confidence limits can be given for the forecast of $\log x$. The occurrence of observations in the extreme tail of the distribution can lead to complicated forms of distortion in estimates of the seasonal component when untransformed data is used. The main consideration in favour of using the multiplicative model in its original form is that its various components have direct meaning in terms of the actual data, whereas this is not the case when we have to transform the data first.

8.1.2 Moving averages

In the study of the models considered in this chapter frequent use is made of the idea of a moving average. We therefore examine this as a technique before looking at its application. In section 4.2 we introduced the idea of a moving average to estimate a local mean. Here we will use the same idea as a technique for manipulating data. The smoothing produced by application of the moving average helps make clearer the underlying variation. Table 4.4 illustrates this point. The original data show an up-and-down movement that tends to hide somewhat the dip in the middle of the year. The moving averages, however, show this feature more clearly. Notice two features in the table:

(a) There are only eight moving averages calculated from the twelve observations. In general, if we calculate a k-valued moving average from m observations, there will be $m - k + 1$ moving averages when k is even and $m - k$ moving averages when k is odd.

(b) If we are to place the moving averages in time, it is reasonable to place them at the middle of the interval covered. In the case where k is even this corresponds to a point in time halfway between two of the times at which data are obtained. Thus if 1 to 12 are the months January to December the first four-month moving average would be halfway between February and March, which may not be very clear in interpretation. A way out of this is to average the adjacent pair of four-month moving averages. The resultant quantity is called a centred moving average. If we do this for the general simple moving average, we finish with $m - k$ centred moving averages.

The importance of the moving average stems from the properties of the series obtained when it is applied to data from certain models. As a convenient notation we will denote by $\{Mx\}$ the series obtained by calculating moving averages from a series, $\{x\}$. The main properties of $\{Mx\}$ are summarized in Table 8.1. Let us comment on these parts

Table 8.1 Effects of application of a moving average M of K terms to the following:

(a) An independent random sequence $\{\epsilon_t\}$, $E(\epsilon) = 0$, $\text{Var}(\epsilon) = \sigma^2$

  (i)   Expectation, $E(M\epsilon) = 0$
  (ii) Variance, $\text{Var}(M\epsilon) = \sigma^2/K$ (approximately $M\epsilon = 0$)
  (iii) Autocorrelation, $\rho(M\epsilon_t, M\epsilon_{t+s}) = 1 - |s|/K$ $(s \leqslant K)$

(b) A linear trend model T, $\alpha + \beta t$ (M centred at t)

  $MT = T$

(c) Pure seasonal model S of K seasons

  $MS = \text{constant}$

---

separately:

(a) If applied to random data the moving average M reduces the variance but leaves the mean unaltered. Thus it can be used to smooth out the random element in data. Unfortunately, a side effect of this smoothing is a high autocorrelation between values of the moving averages. Thus if $K = 12$, the correlation between adjacent values is $11/12$, between next but one values is $10/12$, and so on. High autocorrelations such as these show themselves as a smooth wandering movement in the moving averages, which can be misinterpreted as a genuine oscillation in the series. Appendix A gives a brief discussion of the idea of autocorrelation.

A further aspect of this effect is that the smooth 'oscillation' obtained can be emphasized even more if the sequence of moving averages is again smoothed using another moving average. There is, in fact, a theorem due to Slutsky, 1937, that shows that, under certain conditions, repeated application of moving averages can eventually lead to a sequence that follows a pure sine wave. This result is the exact opposite of what one is usually trying to do when using moving averages. However, this is a 'long run' result and is mentioned simply as a caution and a reminder that one needs to study one's methods carefully before assuming that all is well. In circumstances where one is in doubt about effects such as the above, it is advisable to try out one's methods on simulated data which have known properties and then examine the results.

(b) As is intuitively reasonable, if M is applied to a perfect trend $T = \alpha + \beta t$, the value taken by MT is the value of T at the centre of the moving average. Thus apart from losing the end values M leaves trend data unaltered. If our trend has superimposed error, the same procedure is still reasonable; M simply smooths out this error. There is, however, another valuable way of looking at this. Suppose we fitted the model $\alpha + \beta t$ by least squares to just the K observations used in M. The value of the fitted line at the central point will in fact be the moving average. So the moving average operation corresponds to fitting the trend to a moving section of data and finding the mid-value of the fitted curve. This can clearly be extended by fitting not just linear trends but quadratics, cubics, etc. For example, if we fitted either a quadratic or cubic to a moving section of five observations the fitted value at the middle would be

$$\frac{1}{35}(-3x_{-2} + 12x_{-1} + 17x_0 + 12x_1 - 3x_2)$$

using an obvious notation. Thus the value is of moving average form but with weights attached to the observations. Such weighted moving averages have a wide range of applications in studying seasonal data and in other statistical applications. See Kendall (1973) and Kendall and Stuart (1966) for a detailed discussion of such moving averages. The general conclusion for our purpose is that even if T is not a linear trend we may still be able to find a suitable weighted moving average so that $MT \stackrel{\cdot}{=} T$.

(c) If a moving average with K terms is applied to seasonal data with K 'seasons' in the year, each season will occur once in each calculation of the moving average. Thus, for example, centring M on, say, April will give the same result, for a series with a perfect twelve-month seasonal, as centring it on January when $K = 12$, so a twelve-month moving average will eliminate a seasonal fluctuation of length 12. Thus we can write MS = a constant. We will usually adjust this constant to be zero or one. If this moving average is applied to data showing oscillations that take place in much less than K observations, it will tend to treat them in the same way it treats random fluctuations and smooth them out. If it is applied where the period of oscillation is much greater than K, the moving average will oscillate in the same fashion. If we consider the moving average centred at a peak of such a series, it will contain observations less than the peak value but none greater. Thus the moving average will be lower than the peak value; the more values in M the lower it will be. The converse will happen on the troughs; hence, though the moving average will oscillate with the original data, it will not have as high an amplitude or, putting it another way, it will not penetrate into the peaks and troughs. One way to improve the penetration is to use a weighted moving average where more weight is put on the central observations.

8.1.3 Looking at seasonal data

The analysis of seasonal data has always depended very largely on subjective judgements: initial judgements about the type of model to try, judgements about how often to apply certain smoothing and recalculation operations and judgements towards the end of one's calculations as to how good the assumptions were and whether any slow variations are occurring and should be allowed for. The introduction of more mathematical approaches to seasonal forecasting is reducing and changing the areas of judgement for those who use these methods. However, these approaches do not eliminate the role of judgement. The best basis for any judgement is a careful study of the data and the best way to start this is by an investigation based on graphics and some simple calculations. Figure 8.2 shows various ways of looking at a set of four years of quarterly data. Consider the various plots in turn.

Quarter	1	2	3	4	1	2	3	4	1	2	3	4	1	2	3	4
Data	6	11	17	10	4	10	18	6	1	9	17	5	7	10	16	7

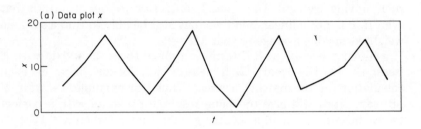

(a) Data plot x

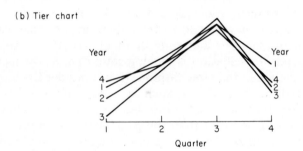

(b) Tier chart

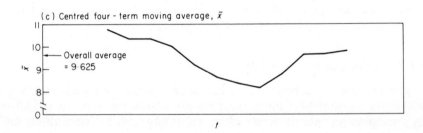

(c) Centred four-term moving average, $\bar{x}$

Overall average = 9·625

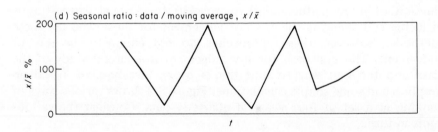

(d) Seasonal ratio : data / moving average , $x/\bar{x}$

(e) Average seasonal ratio's

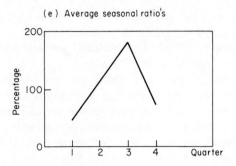

(f) Rough residuals : data – (centred moving average x average seasonal ratio)

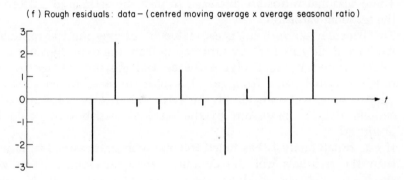

(g) Tier chart of residuals

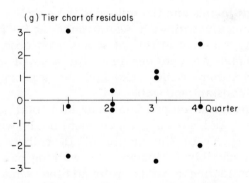

Figure 8.2   Illustrative plots for looking at seasonal data

124

(a) This is a simple plot of the data. Joining the points by line segments often helps to clarify the seasonal pattern.

(b) A tier chart helps to clarify the seasonal pattern and the extent of its variation from year to year.

(c) A centred moving average will show up any major trends. In the example there is no clear trend and the magnitude of the changes are small, suggesting that the level of the series is remaining fairly steady. The very nature of the moving average will produce a fairly smooth curve with a natural wandering movement even when the underlying mean is constant, such as in this example.

(d) A simple seasonal ratio, detailed calculations for which are given in later sections, is plotted against time. Any change in amplitude or in the form of the seasonal pattern, such as illustrated in Figure 8.1(c) and (d), will show up in this plot.

(e) These seasonal ratios are averaged to give the estimated pattern of the seasonal variation.

(f) The irregular component is estimated by calculating the residuals. We have done this here by subtracting from the data the smoothed values, found by multiplying the central moving average by the rough seasonal ratio from (e). The plot of these rough residuals against time will indicate if the error variance is constant; with enough data a histogram of the residual distribution could be produced.

(g) If our rough analysis has failed to cope with the seasonal variation, then the residuals will still contain a seasonal element. This will show up when the residuals are plotted on a tier chart. No such structure is evident from (g).

8.2 Seasonal index methods

8.2.1 Global non-trending models

The methods in this and the following three sections are all based on the idea of associating with each 'season' an index S. Thus S_j measures in some sense the 'seasonal effect' of season j on observations obtained during that time. We consider first, for ease of presentation, the situation where there is no trend and the seasonal pattern can be assumed to be constant and stable.

Let x_{ij} represent an observation made in 'period' i, which might be year i or week i, and 'season' j, which might be the winter quarter $j = 4$ (of $j = 1$, 2, 3, 4) or the month of December, $j = 12$ (of $j = 1$, 2, ..., 12), or possibly a Monday in a working week, $j = 1$ (of $j = 1$, 2, ..., 5). Considering a model with additive random variation ϵ_{ij}, defined in our usual fashion, we could write

$$x_{ij} = \mu_{ij} + \epsilon_{ij} \quad (j = 1, 2, \ldots, r; i = 1, 2, \ldots, t)$$

where μ_{ij} is the 'seasonal mean' for season j in period i. If the seasonal mean is the same or approximately the same from period to period, we could write this as

$$x_{ij} = \mu_j + \epsilon_{ij} \quad (j = 1, 2, \ldots, i = 1, 2, \ldots, t)$$

For each season j this corresponds exactly to the constant mean model of Chapter 4. Hence we can use the formulae of that chapter to provide estimates of μ_j, here denoted by $\mu_{t,j}$. Assuming a global model, we have

$$\hat{\mu}_{t,j} = \sum_{i=1}^{t} x_{ij}/t = \bar{x}_{.j}$$

The layout of the data and the calculation of $\hat{\mu}_{t,j}$ for the global model is illustrated in Table 8.2. The forecast of $x_{t+1,j}$ is simply $\hat{\mu}_{t,j}$.

Table 8.2 Constant seasonal structures — calculations for global model

	Season j		
	1	2	r
Period 1	x_{11}	x_{12}	x_{1r}
2	x_{21}	x_{22}	x_{2r}
t	x_{t1}	x_{t2}	x_{tr}
Totals	$x_{.1}$	$x_{.2}$	$x_{.r}$
$\hat{\mu}_{t,j}$	$\bar{x}_{.1}$	$\bar{x}_{.2}$	$\bar{x}_{.r}$

It is clear that the forecasts of each μ_j are based only on data from that period, so we are really doing separate forecasting exercises using the constant mean model for r different sets of data. For the global model this is obviously the correct thing to do.

Suppose that we wish to refer to the seasonal means μ_j as being relative to a constant mean μ for the whole time of observation. We could do this in two particularly simple ways, either by using an additive form

$$\mu_j = \mu + \phi_j$$

or in a multiplicative form

$$\mu_j = \mu\theta_j$$

So in our notation the trend term T is now a constant mean μ and the

126

seasonal term S_j is given two expressions, ϕ_j and θ_j, according to whether our model is additive or multiplicative. In both cases we would want μ to be the average of the μ_j values, i.e.

$$\mu = \frac{1}{r} \sum_{j=1}^{r} \mu_j$$

and therefore we must have conditions on ϕ_j and θ_j such that

$$\sum_{j=1}^{r} \phi_j = 0 \quad \text{and} \quad \sum_{j=1}^{r} \theta_j = r$$

Table 8.3 illustrates these formulae with some numerical values. It should be emphasized at this point that on the basis of the global model these formulae are just different ways of writing the same thing: μ_j is the fundamental quantity μ, ϕ_j and θ_j are artificially introduced parameters called seasonal indices. When ϕ_j is used, we say that we are using an additive form of model. When θ_j is used, it is called the multiplicative form. We have introduced these artificial forms here as at this stage they are easily understood. When we come to more complicated seasonal models, we will find that these two forms are fundamentally different and not just different ways of writing the same thing. Having introduced the parameters μ, ϕ_j and θ_j, we now ask how to evaluate them. To do this we first write our model in its two new forms:

$$x_{ij} = \mu + \phi_j + \epsilon_{ij}$$

and

$$x_{ij} = \mu\theta_j + \epsilon_{ij}$$

If we average over j for each i, we obtain

$$\bar{x}_{i.} = \mu + \frac{1}{r} \sum_{j=1}^{r} \phi_j + \bar{\epsilon}_{i.}$$

$$\bar{x}_{i.} = \mu \times \frac{1}{r} \sum_{j=1}^{r} \theta_j + \bar{\epsilon}_{i.}$$

which by the conditions on ϕ_j and θ_j gives

$$\bar{x}_{i.} = \mu + \bar{\epsilon}_{i.}$$

in both cases. Taking expectations gives

$$\mu = E(\bar{x}_{i.})$$

so μ is the underlying mean for the average of the data from any period i.

Let us assume for simplicity that there are exactly tr observations, as

Table 8.3 Example of seasonal parameters

Parameter	Jan.	Feb.	Mar.	Apr.	May	June	July	Aug.	Sept.	Oct.	Nov.	Dec.	Mean
μ_j	5	6	7	9	11	13	15	15	13	12	8	6	10
μ	10	10	10	10	10	10	10	10	10	10	10	10	10
ϕ_j	−5	−4	−3	−1	1	3	5	5	3	2	−2	−4	0
θ_j	0.5	0.6	0.7	0.9	1.1	1.3	1.5	1.5	1.5	1.2	0.8	0.6	1.0

in Table 8.2. The average over all seasons, $\bar{x}_{i.}$ $(i = 1, \ldots, t)$, has expectation μ for all periods; thus the average of the $\tilde{x}_{i.}$ over all i will provide the estimate for μ obtained at time t,

$$\hat{\mu}_t = \bar{x}.. = \sum_{i=1}^{t} \bar{x}_i./t$$

which is the 'grand average' of all our data. If instead of averaging over seasons we average over periods for each season, we obtain $\hat{\mu}_{t,j}$ once again. In terms of our two forms of model this is

$$\hat{\mu}_{t,j} = \mu + \phi_j + \hat{\epsilon}._{.j}$$

and

$$\hat{\mu}_{t,j} = \mu\theta_j + \hat{\epsilon}._{.j}$$

where $\hat{\epsilon}._{.j}$ denotes the average of ϵ_{ij} over i. It is now natural to estimate ϕ_j and θ_j by

$$\hat{\phi}_j = \hat{\mu}_{t,j} - \hat{\mu}_t$$

and

$$\hat{\theta}_j = \hat{\mu}_{t,j}/\hat{\mu}_t$$

We thus have reasonable ways of breaking our estimates of μ_j into components $\hat{\mu}$ and $\hat{\phi}_j$ or $\hat{\mu}$ and $\hat{\theta}_j$. A little calculation will show that

$$\sum_{j=1}^{r} \hat{\phi}_j = 0$$

so the additive seasonal indices are positive and negative numbers indicating deviations above and below the mean. Also,

$$\sum_{j=1}^{r} \hat{\theta}_j = r$$

so the multiplicative seasonal indices are numbers above and below one, indicating deviations above and below the mean. The forecast of the value of any further observation $x_{T,j}$ in season j is

$$\hat{x}_{T,j} = \hat{\mu}_t + \hat{\phi}_j$$

or, equivalently,

$$\hat{x}_{T,j} = \hat{\mu}_t\hat{\theta}_j$$

The structure of the above calculation is presented in Table 8.4. If the assumption of exactly tr observations does not hold, it is clearly no use basing a period average on less than r seasons' data since not all values of ϕ_j or θ_j will be included. In this case $\hat{\mu}_t$ could be based on an

Table 8.4 Global and local fitting of models in section 8.2

		Season			Period average	Exponential weights
		1	2	r		
Period	1	x_{11}	x_{12}	x_{1r}	$\bar{x}_{1.}$	a^{t-1}
	2	x_{21}	x_{22}	x_{2r}	$\bar{x}_{2.}$	a^{t-2}
	3	x_{31}	x_{32}	x_{3r}	$\bar{x}_{3.}$	a^{t-3}
	$t-1$	$x_{(t-1)1}$	$x_{(t-1)2}$	$x_{(t-1)r}$	$\bar{x}_{(t-1).}$	a
	t	x_{t1}	x_{t2}	x_{tr}	$\bar{x}_{t.}$	1
Column averages or weighted averages		$\hat{\mu}_{t,1}$	$\hat{\mu}_{t,2}$	$\hat{\mu}_{t,r}$	$\hat{\mu}_t$	
$\hat{\phi}_j$		$\hat{\mu}_{t,1} - \hat{\mu}_t$	$\hat{\mu}_{t,2} - \hat{\mu}_t$	$\hat{\mu}_{t,r} - \hat{\mu}_t$		
$\hat{\theta}_j$		$\hat{\mu}_{t,1}/\hat{\mu}_t$	$\hat{\mu}_{t,2}/\hat{\mu}_t$	$\hat{\mu}_{t,r}/\hat{\mu}_t$		

average of only those periods which contained a full complement of seasons or, alternatively, on an average of the r seasonal averages. These two estimates of μ will not be the same, as the latter will use all the data and the former only part of the data.

8.2.2 Local non-trending models

We now turn to the situation where we feel the need to consider fitting our model locally rather than globally. In this context this means that we will consider methods that allow for slow changes in the seasonal indexes as new data arrive. Remembering that our basic model is

$$x_{ij} = \mu_j + \epsilon_{ij} \quad (j = 1, 2, \ldots, r; i = 1, 2, \ldots, t)$$

which is the ordinary constant mean model for each j, an estimate of μ_j during the most recent period t, using exponential weights, is

$$\hat{\mu}_{t,j} = \sum_{i=1}^{t} a^{t-i} x_{ij} \bigg/ \sum_{i=1}^{t} a^{t-i}$$

This is exactly as discussed in Chapter 4.

To estimate the underlying mean μ, a number of estimates could be considered. Of these the weighted mean of the ordinary 'yearly' means will be used, i.e.

$$\hat{\mu}_t = \sum_{i=1}^{t} a^{t-i} \bar{x}_{i.} \bigg/ \sum_{i=1}^{t} a^{t-i}$$

This form has the advantage that if we define $\hat{\phi}_j$ and $\hat{\theta}_j$, as before, by

$$\hat{\phi}_j = \hat{\mu}_{t,j} - \hat{\mu}_t$$

and

$$\hat{\theta}_j = \hat{\mu}_{t,j}/\hat{\mu}_t$$

then all the results of section 8.2.1 still hold for the local model. We also have, for large t, the recurrence relations

$$\hat{\mu}_{t,j} = (1 - a)x_{tj} + a\hat{\mu}_{t-1,j}$$

and

$$\hat{\mu}_t = (1 - a)\bar{x}_{t.} + a\hat{\mu}_{t-1}$$

The above equations provide an 'exact' type of solution together with a recurrence form for large t.

Let us consider now the estimation of the parameters μ, ϕ_j and θ_j in a purely intuitive fashion and with the object of expressing our estimates in the form of recurrence relations, such that $\hat{\mu}_t$ can be updated more frequently than once a year. We thus want something in the form

New estimate = constant x new information + constant x old estimate

We have already an expression for $\hat{\mu}_{t,j}$ in this form. If we wish to get this form for the seasonal indexes ϕ_j and θ_j, we could subtract or divide by the latest estimate, $\hat{\mu}_t$, of the underlying mean. Subtracting $\hat{\mu}_t$ from both sides of the expression for $\hat{\mu}_{t,j}$ gives

$$\hat{\mu}_{t,j} - \hat{\mu}_t = (1 - a)(x_{tj} - \hat{\mu}_t) + a(\hat{\mu}_{t-1,j} - \hat{\mu}_t)$$

To make the use of this expression a practical proposition, we must rewrite it in an approximate form, namely

$$\hat{\mu}_{t,j} - \hat{\mu}_t = (1 - a)(x_{tj} - \hat{\mu}_t) + a(\hat{\mu}_{t-1,j} - \hat{\mu}_{t-1})$$

where in the last term we have had to subtract $\hat{\mu}_{t-1}$ instead of $\hat{\mu}_t$ to correspond to the fact that at time $t-1$ the estimate of the underlying mean was $\hat{\mu}_{t-1}$. In terms of $\hat{\phi}$ this is

$$\hat{\phi}_{t,j} = (1 - a)(x_{tj} - \hat{\mu}_t) + a\hat{\phi}_{t-1,j}$$

Similarly, for $\hat{\theta}$ we would have

$$\hat{\theta}_{t,j} = (1 - a)(x_{tj}/\hat{\mu}_t) + a\hat{\theta}_{t-1,j}$$

Using these we can update estimates of the seasonal factors ϕ and θ provided $\hat{\mu}_t$ is known. We now want to replace $\hat{\mu}_t$ by a form in which we can update the estimate of μ using each observation as it is obtained. Let us denote such an estimate made in season j of period t by $\hat{\mu}_{(t,j)}$. A reasonable recurrence relation for $\hat{\mu}_{(t,j)}$ would take the form

$$\hat{\mu}_{(t,j)} = (1 - b)(\text{latest information about } \mu) + b\hat{\mu}_{(t,j-1)}$$

where b is again a constant $0 < b < 1$. What to use as 'the latest information about μ' is suggested by our basic model, since using only the one observation and estimates of ϕ and θ, μ is estimated by $x_{tj} - \hat{\phi}_j$ and $x_{tj}/\hat{\phi}_j$. The recurrence equations thus become

$$\hat{\mu}_{(t,j)} = (1 - b)(x_{tj} - \hat{\phi}_{t-1,j}) + b\hat{\mu}_{(t,j-1)}$$

and

$$\hat{\mu}_{(t,j)} = (1 - b)x_{tj}/\hat{\theta}_{t-1,j} + b\hat{\mu}_{(t,j-1)}$$

where in both cases we have used estimates of ϕ and θ obtained at time $t - 1$, since the estimates at time t cannot be obtained until $\hat{\mu}_{(t,j)}$ is known. Having obtained a recurrence relation for $\hat{\mu}_{(t,j)}$ we may use this in place of $\hat{\mu}_t$ in the recurrence relations for $\hat{\phi}_j$ and $\hat{\theta}_j$. The quantities $x_{t,j} - \hat{\phi}_{t-1,j}$ and $x_{t,j}/\hat{\theta}_{t-1,j}$ are termed the seasonally adjusted observations, since they are essentially our estimates of what the observations would have been had there been no seasonal effects. As is usual with recurrence relations, we can re-express them in error-correction form. Thus denoting the one-step-ahead forecast error at season j of period t by $e_{t,j}$, we have

$$e_{t,j} = x_{t,j} - (\hat{\mu}_{(t,j-1)} + \hat{\phi}_{t-1,j})$$

or

$$e_{t,j} = x_{t,j} - (\hat{\mu}_{(t,j-1)}\hat{\theta}_{t-1,j})$$

Then we may rewrite both the recurrence equations as

$$\hat{\mu}_{(t,j)} = \hat{\mu}_{(t,j-1)} + (1-b)e_{t,j}$$

The sequence of calculations is as indicated in Figure 8.3.

As on previous occasions we will postpone a discussion of the choice of forecasting parameters a and b until a later chapter. However, it should be noted that b should be very close to one, otherwise the natural variation in $\hat{\mu}_{(t,j)}$ will distort the estimates $\hat{\phi}$ and $\hat{\theta}$ and the process of forecasting will tend to lead to erratic results. It might in fact be safest to use previous methods discussed for calculating $\hat{\mu}_t$ at the end of each period and to use this value as the initial value in calculating the $\hat{\mu}_{(t+1,j)}$ values for the r seasons in the following period.

In the global model situation we noted that the difference between additive and multiplicative forms was purely arbitrary and the answers obtained were identical. In the recurrence relations above this is no longer the case. In a local situation the underlying mean will wander. There is thus a real difference between the situation where an additive

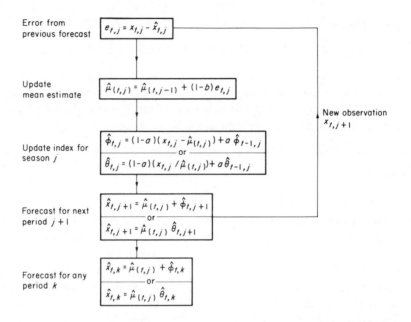

Error from previous forecast	$e_{t,j} = x_{t,j} - \hat{x}_{t,j}$
Update mean estimate	$\hat{\mu}_{(t,j)} = \hat{\mu}_{(t,j-1)} + (1-b)e_{t,j}$
Update index for season j	$\hat{\phi}_{t,j} = (1-a)(x_{t,j} - \hat{\mu}_{(t,j)}) + a\,\hat{\phi}_{t-1,j}$ or $\hat{\theta}_{t,j} = (1-a)(x_{t,j}/\hat{\mu}_{(t,j)}) + a\hat{\theta}_{t-1,j}$
Forecast for next period $j+1$	$\hat{x}_{t,j+1} = \hat{\mu}_{(t,j)} + \hat{\phi}_{t,j+1}$ or $\hat{x}_{t,j+1} = \hat{\mu}_{(t,j)}\,\hat{\theta}_{t,j+1}$
Forecast for any period k	$\hat{x}_{t,k} = \hat{\mu}_{(t,j)} + \hat{\phi}_{t,k}$ or $\hat{x}_{t,k} = \hat{\mu}_{(t,j)}\,\hat{\theta}_{t,k}$

New observation $x_{t,j+1}$

Figure 8.3   Recursive calculation of the mean and seasonal factors

seasonal ϕ adds the same amount on to μ_t, whatever its value, and the multiplicative situation where an increase in μ_t will increase the total variation in the seasonal pattern represented by $\theta\mu_t$.

In the above we have assumed that the seasonal element, $S = \theta$ or $S = \phi$, wanders in a slow and irregular fashion. There may, however, be systematic change taking place in S. This possibility should be examined by looking at the variation of S with time. The quantities

$$\hat{S}_{ij} = x_{ij} - \hat{\mu}_t \text{ and } \hat{S}_{ij} = x_{ij}/\hat{\mu}_t$$

give, for additive and multiplicative models respectively, an estimate of the value of S at time t, which as before in terms of periods and seasons is (i,j). A plot of S_{ij} as a time series will give a general indication of any systematic change taking place. More detail can be found by looking, separately, for each season; how S_{ii} is changing with i. If there is a clear increase or decrease in S_{ij} with i for any reason j, a forecast of future values can be obtained by extrapolating the trend.

8.2.3 Models with trend

In the previous sections it helped the presentation to distinguish between global and local approaches to fitting and extrapolating the seasonal model. When we turn now to seasonal models with a trend component, the clarity of distinction gets lost when seasonal index methods are used. Basically the methods are designed for use in situations where the trend component is unlikely to be a simple mathematical trend and thus the whole approach tends to look at the situation in 'local' terms. There are a number of points in the procedure where, if the situation is thought to be stable, an ordinary average might be used or a mathematical curve fitted. If this stability is doubted, exponential smoothing might be used instead. The general procedures, however, are common and it is these that we will study.

Where a trend is present any attempt to derive seasonal components using the methods of the previous section will fail because the seasonals would contain contributions from the trend. To find the seasonals we must first find the trend and remove its contribution. There are two elementary methods of doing this. One is to find an empirical trend curve by a suitable smoothing method. The other is to fit a mathematical trend curve. Let us examine these briefly in turn.

(a) Fitting an empirical trend curve

The basic method of obtaining a trend for data of seasonal period r is to use a simple moving average of r terms. To see the justification for this consider the additive model

$$x_{ij} = T_{ij} + S_j + I_{ij}$$

Applying the moving average operation centred at time i,j gives

$$Mx_{ij} = MT_{ij} + MS_j + MI_{ij}$$

Using the results of the section on moving averages

$$MT_{ij} = T_{ij}$$

$$MS_j = \text{constant} = 0 \text{ by choice}$$

$$MI_{ij} = 0 (\text{approximately})$$

Hence

$$Mx_{ij} \doteq T_{ij}$$

and the sequence of values given by the moving averages approximately follows the underlying trend at the centres of the moving averages. If the model is multiplicative, we will get $M(T_{ij}S_j)$, which does not conveniently simplify unless T_{ij} is nearly constant over the time spanned by the moving average. We might in this case convert to an additive model by taking logs of the data and then converting back the estimated value of T_{ij} obtained.

(b) Fitting a mathematical trend curve

Before we can fit a trend curve we first need to get rid, at least approximately, of the seasonal component. This can be done most simply by using a period average in which the seasonal component is summed out, as it is in the moving average. These averages can now be fitted by an appropriate mathematical curve using the method of least squares or any other appropriate method, we have already discussed fitting straight lines in Chapter 5 and we will discuss fitting non-linear curves in a later chapter. However, there is a complication that requires a brief mention, and that is that the period averages do not lie on the same curve as the original data. For example, if monthly data showing slope β are used to get the averages for a sequence of years, these yearly averages will increase by 12β from year to year. However, if the original data, the averaged data and fitted lines are plotted together, there should be no confusion in obtaining the correct estimates.

Both the above methods are open to criticism and the choice between them depends somewhat on how appropriate these criticisms are in one's own particular circumstances. The main aspects to be considered are as follows.

(a) Empirical curves

The main criticism of these curves *is* that they are empirical. They do not have a specified mathematical structure. This fact has two unfortunate consequences. Firstly, as we do not have a structure defined, we cannot easily extrapolate into the future to obtain our forecasts. Secondly, the lack of a model makes almost impossible an examination of how well the method has determined the past trend. The method really provides one definition of what is meant by the trend.

(b) Mathematical curves

If a very simple mathematical model such as a linear trend is used, it is usually too simple to describe any realistic trend and will certainly not include the cycle. However, if a much more complicated model is used, it may well tend to pick up parts of the seasonal structure as well as the cyclic structure.

We started this section by saying that before we could estimate the seasonal indexes S_j we had first to find and eliminate the effect of the trend component T_{ij}. We have now found methods of obtaining at least a rough estimate $\hat{T}_{ij}$ of this, so let us now continue by finding an estimate $\hat{S}_j$ of S_j. Assuming still an additive model we can get a first estimate of S_j, using observation x_{ij}, by subtracting $\hat{T}_{ij}$, giving

$$\hat{S}_{ij} = x_{ij} - \hat{T}_{ij}$$

From the model this quantity will be seen to be approximately

$$S_j + I_{ij}$$

To estimate S_j then we must now reduce the effect of I_{ij} by averaging the above quantities over i,

$$\hat{S}_j = \text{average of } \hat{S}_{ij} \quad (j = 1, \ldots, r)$$

The sequence of steps is thus:

(a) Estimate T_{ij} by a moving average or fitting procedure, giving $\hat{T}_{ij}$.
(b) Subtract these from the original data, $x_{ij} - \hat{T}_{ij}$.
(c) Average over all periods for each season.

We now have estimates of both T_{ij} and S_j so we might go ahead and produce a forecast. However, our methods for finding the estimate of T_{ij} left a lot to be desired, so in practice we seek to improve our estimate first, As we now have an estimate $\hat{S}_j$ of S_j we can seek to remove the seasonal effect by division or subtraction to get seasonally adjusted data

$$x'_{ij} = x_{ij}/\hat{S}_j \quad \text{or} \quad x'_{ij} = x_{ij} - \hat{S}_j$$

If the seasonal was found in the previous investigations to vary in some fashion, then $\hat{S}_j$ here would be replaced by our best estimate of S at time i,j.

The aim of the next few pages is to discuss again the estimation of T_{ij}. However, we now have seasonally adjusted data as the basis for our calculations rather than the original data. With the influence of S_j at least approximately removed, the object is now to seek to estimate T_{ij} by reducing the effect of the irregular component I_{ij}. This may be done by fitting by eye, by fitting some empirical trend, using some type of moving average technique, or by fitting a mathematical trend curve. Considering the latter two in turn.

(a) Moving average techniques

A natural intuitive method of reducing the effect of the irregular component is to smooth it out using a moving average of some form. A traditional way of doing this is to use a weighted moving average to fit a polynomial to a section of data. For example, suppose we choose a section of five observations, called x'_{-2}, x'_{-1}, x'_0, x'_1 and x'_2, and fit a linear trend by least squares. The simple moving average is just the average of the five observations. Now we have seen before that the trend line goes through the mean of the observations, so the value of the fitted line at the time corresponding to x_0' is

$$\frac{1}{5}(x_{-2}' + x_{-1}' + x_0' + x_1' + x_2')$$

often denoted by

$$\frac{1}{5}[1, 1, \underline{1}]$$

making use of the symmetry and listing only the coefficients of x'. If we had fitted either a quadratic or a cubic, the value of the fitted curve at the midpoints would have been

$$\frac{1}{35}(-3\,x_{-2}' + 12\,x_{-1}' + 17\,x_0' + 12\,x_1' - 3\,x_2')$$

or

$$\frac{1}{35}[-3, 12, \underline{17}]$$

Thus as the above five-pointed weight moving average is moved through the data, it effectively fits a quadratic or cubic (the result is the same) to the successive sections of data and gives the midpoints of the fitted curves.

One disadvantage of using the various types of moving average is that they estimate only the trend at the middle of the section used. When a $2\,m + 1$ moving average is moved through a set of data, no trend value is estimated for the first and last m values. If one is using a weighted moving average to fit a curve by least squares, then one can give the values of the fitted curve over the m end values and indeed extrapolate past these. A danger here is that only a relatively small number of observations are used in the estimation, so that extrapolation of, say, a cubic to the end of the moving average can lead to estimates with large variances. An approach sometimes used, though of debatable logic from the forecasting viewpoint, is to forge extra m observations at either end of the data being used. An inappropriate choice of forged values will affect $2\,m$ of the moving averages and possibly lead to incorrect interpretations and forecasts.

An alternative approach here is to fit a trend curve using the two-way exponential weights (see Appendix C). This enables one to obtain estimates at all points of the data, for the maximum weights can be placed successively at each point in the data.

The great disadvantage of this approach is that it produces an empirical trend curve over past data which has no natural way of extrapolating into the future. It does, however, provide a means of analysing past data as a preliminary to forecasting.

(b) Fitting a curve

As an alternative to the above we may seek to fit, perhaps by discounted least squares, a simple mathematical curve to the seasonally adjusted sequence. For forecasting purposes this is a sensible procedure as we are now provided with a means of extrapolating the trend into the future. Consider the two simplest cases of local constant mean and trend models. In the former we could update the mean now given by T_{ij} using exponential smoothing. Thus

$$\hat{T}_{ij} = a\hat{T}_{i,j-1} + (1-a)x'_{ij}$$

where $\hat{T}_{i,j-1}$ denotes the last estimated trend value, ignoring whether it was in period i or not, and x'_{ij} denotes the latest seasonally adjusted observation. At this stage only $\hat{S}_{i-1,j}$ is known, so

$$x'_{ij} = x_{ij}/\hat{S}_{i-1,j}, \text{ for the multiplicative (x) model}$$

or

$$x'_{ij} = x_{ij} - \hat{S}_{i-1,j}, \text{ for the additive (+) model}$$

The natural formula for obtaining the $\hat{S}_{i,j}$ is that given at the end of the previous section, namely

$$\hat{S}_{i,j} = b\hat{S}_{i-1,j} + (1-b)S_{ij}$$

where the simplest form for S_{ij} is

$$S_{ij} = x_{ij}/\hat{T}_{ij}, \text{ for the multiplicative model}$$

or

$$S_{ij} = x_{ij} - \hat{T}_{ij}, \text{ for the additive model}$$

The methods of (a) above and of the previous sections can be used on an initial set of data to provide the values from which to start the recurrence relations. The sequence of the recurrence relation is given in Figure 8.4(a).

The forecast of a future value $x_{r,s}$ will be

$$\hat{x}_{r,s} = \hat{T}_{ij} + \hat{S}_{is} \text{ (+ model) or } \hat{T}_{ij}\hat{S}_{is} \text{ (x model)}$$

where $\hat{T}_{ij}$ is the latest trend estimate and $\hat{S}_{is}$ is the latest seasonal factor for season s. If a study of past data can give good starting values for the recurrence relations, it is advisable to keep the two forecasting parameters a and b quite close to 1.0 otherwise such a set of equations

(a) Locally constant mean form

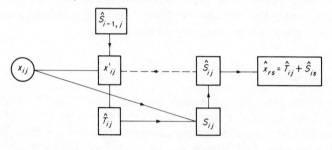

(b) Local trend form

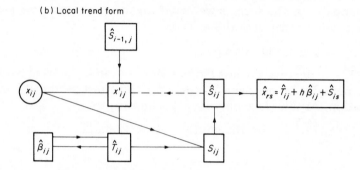

Figure 8.4   Sequence of calculations for recursive forecasting

can easily lead to wildly varying values. If there is evidence of a change in mean and one wishes to pick this up rapidly, a would be reduced but not b — conversely, if a large change in seasonal pattern is suspected.

If the model is a local linear trend, we would need recurrence relations for the current level of the trend T_{ij} and for the slope β_{ij}. The natural relations to use are those of Chapter 5; thus

$$\hat{T}_{ij} = a(\hat{T}_{i,j-1} + \hat{\beta}_{i,j-1}) + (1-a)x'_{i,j}$$

$$\hat{\beta}_{ij} = c\hat{\beta}_{i,j-1} + (1-c)(\hat{T}_{ij} - \hat{T}_{i,j-1})$$

The recurrence relation for $\hat{S}_{ij}$ will be as before. A forecast of an observation x_{rs} which is h units into the future is

$$\hat{x}_{rs} = (\hat{T}_{ij} + h\hat{\beta}_{ij}) + \hat{S}_{is}$$

The sequence of calculations here is given in Figure 8.4(b) and is often called the Holt—Winter's method.

A slight modification of this procedure is based on the observation that x'_{ij}, the seasonally adjusted observation, is based on the use of the last seasonal factor $\hat{S}_{i-1,j}$. Having gone once through the sequence of calculations in Figure 8.4(a) and (b) we can use the new value $\hat{S}_{i,j}$ to revise x'_{ij} and iterate the calculation to obtain what are, intuitively, better values for $\hat{T}_{ij}$, $\hat{S}_{ij}$ and $\hat{\beta}_{ij}$.

The literature on seasonal methods gives extensive study to ways of cycling round these various calculations seeking at each stage to improve these estimates. Durbin and Murphy (1975) and Harrison (1965) discuss aspects of these calculations in detail. Basically, the modifications are concerned with:

(a) reducing the effects of extreme observations, possibly caused by known factors;
(b) adjusting the seasonal estimates so that they satisfy the conditions that $\Sigma\hat{S} = 0$ or $\Sigma\hat{S} = 12$;
(c) introducing more sophisticated forms of moving average to deal with particular features of the data.

8.3 Fourier methods

Over the years many criticisms have been made of the iterative approach described in the previous sections, particularly where it is applied to the multiplicative model. For example, Lovell (1963) and Jorgenson (1967) have stated criteria that a good method for fitting seasonal models should satisfy; in general, the methods discussed above satisfy neither of these sets of criteria.

Particularly relevant criticisms from the forecasting viewpoint are

(a) the lack of any estimates of the error variances and hence of prediction errors. The only way round this is to estimate these by applying the methods to past data.
(b) the lack, in some formulations of the approach, of a mathematical model that can be extrapolated.

In spite of these criticisms we have examined these methods at some length as they do provide a robust approach that has led to several simple methods of forecasting. Our study has also enabled us to introduce a number of ideas of importance in other approaches to the analysis of seasonal models. The aim of this section and the following section is to give a brief survey of some of the alternative approaches.

The essence of most of the recent methods of seasonal analysis is to express the model in additive form and to apply to such a linear form of model a standard method of estimation. Thus the basic model for the observations, or their logarithms, is

$$x_{ij} = T_{ij} + S_{ij} + I_{ij}$$

where each of the three parts can be modelled in some sensible way. For example, the trend might be modelled by a polynomial in t (see, for example, Duvall, 1966). The irregular component might be modelled by a set of independent normal or log-normal variables or by some simple stochastic process. The main focus of recent work however, has been to model the seasonal component using modifications of methods that date back to the last century.

The basis of the methods is to fit the seasonal component with a regression model in which the regressor variables are periodic functions of time. Suppose, for example, that we have monthly data with twelve seasonal indices, $S_1, \ldots, S_{12}$. The methods called Fourier analysis show that this sequence of twelve points lie on the curve

$$S_j = 1 + \sum_{k=1}^{6} \left(u_k \cos \frac{2\pi\, kj}{6} + v_k \sin \frac{2\pi\, kj}{6} \right)$$

This is a regression curve with regression coefficients u_k and v_k given by

$$u_k = \frac{1}{6} \sum_{j=1}^{12} S_j \cos \frac{2\pi\, kj}{6} \qquad (k = 1, \ldots, 6)$$

and

$$v_k = \frac{1}{6} \sum_{j=1}^{12} S_j \sin \frac{2\pi\, kj}{6} \qquad (k = 1, \ldots, 5)$$

$$v_6 = 0$$

Thus we have converted from one set of twelve indices with the additive condition

$$\left(\sum_{j=1}^{12} S_j = 12 \right)$$

to another equivalent set of eleven regression coefficents. Thus at this level there is no great merit in introducing such a Fourier model. The advantage of these models comes from the practical observation that one usually obtains a good fit without having to use all twelve terms in the regression model. Any pair of sin and cos terms, say,

$$u_k \cos \frac{2\pi\, kj}{6} + v_k \sin \frac{2\pi\, k}{6}\, j$$

represents a pure oscillation with amplitude

$$a_k = \sqrt{(u_k^2 + v_k^2)}$$

It may be observed, for example, that a_1 and a_2 are large but $a_3 \ldots a_6$ are relatively small. The procedure then is to use the model

$$S_j = 1 + u_1 \cos \frac{2\pi\, j}{6} + v_1 \sin \frac{2\pi\, j}{6} + u_2 \cos \frac{4\pi\, j}{6} + v_2 \sin \frac{4\pi j}{6}$$

The coefficients u_1 v_1, u_2 and v_2 can be found by fitting this as a linear regression model to the latest seasonal indexes obtained by the methods of the previous section. Such a method of replacing the usually rather irregular $\hat{S}_j$ values by a smoother form of seasonal variation is referred to as 'Fourier smoothing' (Harrison, 1965). In general this model would have to be fitted each time a new observation and estimate of S was obtained. However, if each new $\hat{S}$ is obtained by a recurrence relation from the previous value for that period, as discussed in the previous section, then recurrence relations can be obtained for the coefficients u_1, v_1, etc., and for the smoothed seasonals themselves.

Thus we have

$$\hat{S}_{i,j} = \hat{S}_{i-1,j} + (1-b)(S_{ij} - \hat{S}_{i-1,j})$$

which is the 'error correction' form of the recurrence relation of the previous section. Let $\tilde{S}_{j,t}$ denote the smoothed seasonal obtained at time t for season j. Then

$$\tilde{S}_{k,t} = \tilde{S}_{k,t-1} + \text{constant} \times (S_{ij} - \tilde{S}_{j,t-1})$$

Note that as each new S_{ij} is obtained we fit the curve over all seasons. Thus we update all seasons, k, on the basis of the new information about season j, S_{ij} and the previous information about season k, $\tilde{S}_{k,t-1}$. The 'constant' does in fact depend on j and k but not on the data. It can be written as

$$\frac{2(1-b)}{n} \sum_r \cos \frac{2r(k-1)}{n}$$

where n is the number of seasons and the summation is over the number of sin, cos pairs used in the Fourier model; in our example this is two.

A number of variations on this theme are discussed in the paper by Harrison. These methods have a number of advantages. They combine a mathematical model for the seasonal component with some of the standard methods for dealing with trends. This enables us in effect to reduce the number of parameters that we are using. By use of different smoothing constants for the seasonal and the trend fitting the methods can allow for changing trends in a fixed seasonal pattern and vice versa. The use of the fitted Fourier terms that are dependent on all the seasonals also fits in with one's intuitive feeling that in most situations the seasonal components are not independent of each other with one able to vary without effecting those on either side. The Fourier approach links the seasonals together in an intuitively reasonable fashion.

A natural extension of the above approach is to modify it to deal with the possibility of changing seasonal patterns. Whether this is the case can be investigated by plots of the u and v values and amplitudes as they are recalculated to allow for more data. One natural extension is to make u and v linear functions of time; alternatively, the whole seasonal expression can be multiplied by a time-varying term. The work by Nettheim (1964) and Duval (1966) deals with fitting these forms of model. In such applications we usually have a linear regression model and so the classical approach is to fit by least squares. We have already seen that discounted least squares with exponential weights a^r reduces to ordinary least squares when $a = 1$. We will therefore look at the fitting by discounted least squares rather than the special case of ordinary least squares. If we seek to reduce our problem to a linear regression form of model, it is natural to include the trend in the model and not just the seasonal, as we did above. Examples of linear models of use in seasonal forecasting are

$$x_t = \beta_0 + \beta_1 t + \beta_2 \sin \frac{2\pi t}{n} + \beta_3 \cos \frac{2\pi t}{n} + \epsilon_t$$

$$x_t = \beta_0 + (\beta_2 + \gamma_2 t)\sin \frac{2\pi t}{n} + (\beta_2 + \gamma_3 t)\cos \frac{2\pi t}{n} + \epsilon_t$$

$$x_t = \beta_0 + \beta_2 \sin \frac{2\pi t}{n} + \beta_3 \cos \frac{2\pi t}{n} + \beta_4 \sin \frac{4\pi t}{nn} + \beta_5 \cos \frac{4\pi t}{n} + \epsilon_t$$

In the first example an additive linear trend is included with the simple seasonal. This has clear advantages in terms of forecasting when the situation is fairly stable. If, however, this is not the case and independent changes can occur in the trend levels and the seasonal patterns of the data, then the previous methods of separately handling T and S are likely to give better results. When we fit by discounted least squares the smoothing constant a is clearly applied to the whole model, and having different smoothing constants for the trend and seasonal parts is clearly ruled out. In the second example the amplitude and position of the simple seasonal pattern is changing in time. In the last example a more complicated seasonal pattern is allowed for by including more Fourier terms. As a means of indicating some of the procedures used in forecasting from such models, let us consider in a little more detail how we could locally fit a simple linear model with a trend and seasonal component. Suppose the fitted model takes the form

$$x_t = b_0 + b_1 t + b_2 \sin \frac{2\pi t}{n} + b_3 \cos \frac{2\pi t}{n} + \hat{\epsilon}_t$$

where $\hat{\epsilon}_t$ is the residual and b_0, b_1, b_2 and b_3 are discounted least squares coefficients chosen to minimize

$$S = \sum_{r=0}^{t} a^r \hat{\epsilon}_{t-r}$$

This is simply a linear regression model with 1, t, $\sin 2\pi t/n$ and $\cos 2\pi t/n$ acting as regressor variables. As usual the values of b_0, b_1, b_2 and b_3 are obtained by solving the set of normal equations. With a slight alteration of the notation of section 6.2, the four normal equations are

$$b_0 S(1,1) + b_1 S(1,t) + b_2 S(1,s) + b_3 S(1,c) = S(1,x)$$

$$b_0 S(t,1) + b_1 S(t,t) + b_2 S(t,s) + b_3 S(t,c) = S(t,x)$$

$$b_0 S(s,1) + b_1 S(s,t) + b_2 S(s,s) + b_3 S(s,c) = S(s,x)$$

$$b_0 S(c,1) + b_1 S(c,t) + b_2 S(c,s) + b_3 S(c,c) = S(c,x)$$

where $S(\ )$ is defined by the formula

$$S(z,v) = \sum_{r=0}^{t} a^r z_{t-r} v_{t-T}$$

Thus, for example,

$$S(s,t) = \sum_{r=0}^{t} a^r \sin \frac{2\pi(t-r)}{n} (t-r)$$

and

$$S(c,x) = \sum_{r=0}^{t} a^r x_{t-r} \cos \frac{2\pi(t-r)}{n}$$

In practice we have a large, but straightforward, exercise in calculation to evaluate all the coefficients, $S(\ )$, and to solve the normal equations. The forecast is obtained as usual by setting the future error term at zero and substituting the future time $t + h$ in the fitted model. Hence

$$\hat{x}_{t+h} = b_0 + b_1(t+h) + b_2 \sin \frac{2\pi(t+h)}{n} + b_2 \cos \frac{2\pi(t+h)}{n}$$

The above linear regression approach can clearly be extended to include more terms in the model. Often more terms from the Fourier series are included, in particular one would usually include $\sin 4\pi\, t/n$, $\cos 4\pi\, t/n$ and possibly $\sin 6\pi\, t/n$, $\cos 6\pi\, t/n$. It is also sometimes useful to include terms that allow for a trend in the amplitude of the oscillations, e.g. $t \sin 2\pi\, t/n$ and $t \cos 2\pi\, t/n$.

The above formal solution of the normal equations involves a considerable amount of computation. when it has to be repeated each time a new observation is obtained, the work can get too extensive for practical use. As a consequence of this, techniques have been introduced that reduce the amount of calculation substantially. There are three main elements in these techniques.

(a) The introduction and evaluation of new trigonometrical functions required for updating the S quantities can be avoided by the use of trigonometrical relations. Thus when x_{t+1} is introduced, we would require to know $\sin 2\pi(t+1)/n$ and $\cos 2\pi(t+1)/n$. However, these are related to the previous values, $\sin 2\pi t/n$ and $\cos 2\pi t/n$, by standard trigonometry:

$$\sin \frac{2\pi(t+1)}{n} = \sin \frac{2\pi t}{n} \cos \frac{2\pi}{n} + \cos \frac{2\pi t}{n} \sin \frac{2\pi}{n}$$

$$\cos \frac{2\pi(t+1)}{n} = -\sin \frac{2\pi t}{n} \sin \frac{2\pi}{n} + \cos \frac{2\pi t}{n} \cos \frac{2\pi}{n}$$

Thus the evaluation on the computer of trigonometric functions can be replaced by simple multiplication and addition, using the constants $\sin 2\pi/n$ and $\cos 2\pi/n$.

(b) The second technique is one that we have already made use of in our discussion of the linear trend model, namely the constant shifting of the origin to the present. With the trend we started with

$$x_t = \alpha' + \beta t$$

but then rewrote the model as

$$x_{t-r} = \alpha_t - \beta r$$

so that α_t always referred to the mean of the trend line at the time of the latest observation. This requires the updating of the meaning of the parameters. Here, for example,

$$\alpha_{t+1} = \alpha_t + \beta$$

This has the advantage that the forecast for a lead time h using estimates $\hat{\alpha}_t$ and $\hat{\beta}_t$ always takes the form

$$\hat{x}_{t+h} = \hat{\alpha}_t + \hat{\beta}_t h$$

To illustrate this let us take the mean and trend terms to be zero and consider $b_{2,t}$ and $b_{3,t}$ as the fitted values for the model which takes the latest observation time, t on the absolute scale, as the origin. Thus the forecast corresponding to the above trend would be

$$x_{t+h} = b_{2,t} \sin \frac{2\pi h}{n} + b_{3,t} \cos \frac{2\pi h}{n}$$

We will clearly wish to update the model parameters so that at time $t+1$ we will know the relation of the new parameters to the previous values at time t. This can be done by making use of the trigonometrical formulae used in (a) above.

Considering the underlying model, with parameters $\beta_{2,t}$ and $\beta_{3,t}$ for origin at time t. If we write the underlying mean value of the series for time $t + h + 1$, treating time $t + 1$ as origin, we have

$$\beta_{2,t+1} \sin \frac{2\pi h}{n} + \beta_{3,t+1} \cos \frac{2\pi h}{n}$$

If, however, we write down the same value using t as the origin, it takes the form

$$\beta_{2,t} \sin \frac{2\pi(h + 1)}{n} + \beta_{3,t} \cos \frac{2\pi(h + 1)}{n}$$

Using the same formulae as in (a), this last expression can be put in terms of $\sin 2\pi h/n$ and $\cos 2\pi h/n$. Equating the coefficients of $\sin 2\pi h/n$ and $\cos 2\pi h/n$ in the two expressions gives

$$\beta_{2,t+1} = \beta_{2,t} \cos \frac{2\pi}{n} - \beta_{3,t} \sin \frac{2\pi}{n}$$

and

$$\beta_{3,t+1} = \beta_{2,t} \sin \frac{2\pi}{n} + \beta_{3,t} \cos \frac{2\pi}{n}$$

Thus we have again a simple way of updating the model coefficients so that we can use as a model one that takes the latest observation time as its working origin. Taking the present time t as origin, the 'S' terms in the normal equations all take the form

$$S_t(z,v) = \sum_{r=0}^{t} a^r z_{-r} v_{-r}$$

So, for example,

$$S_t(s,t) = \sum_{r=0}^{t} ra^r \sin \frac{2\pi r}{n} \qquad \text{(the two } - \text{ signs giving a +)}$$

and

$$S_t(c,x) = \sum_{r=0}^{t} a^r x_{t-r} \cos \frac{2\pi r}{n}$$

Solving the appropriate normal equations in this form at any time t will give fitted values $b_{2,t}$ and $b_{3,t}$.

(c) The results in (b) give a way of updating the model parameters with shift of origin. However, in updating the fitted parameters we need to introduce into the situation the latest observation. To deal with this aspect we can use the fact that all the $s(\ )$ terms in the normal equations can be updated in a simple fashion:

$$S_{t+1}(z,v) = z_{t+1} v_{t+1} + aS_t(z,v)$$

This is an exact relation and can be generally used for updating.

This recurrence relation would always be used for updating the terms on the right-hand side of the normal equations since these will involve the introduction of the new observation x_{t+1}. If we use the present as the origin and have a large amount of data, we can get an approximate result that simplifies calculation considerably. In this form, from (b), we have

$$S_{t+1}(z,v) = \sum_{r=0}^{t+1} a^r z_{-r} v_{-r}$$

so

$$S_{t+1}(z,v) = S_t(z,v) + a^{t+1} z_{-(t+1)} v_{-(t+1)}$$

For all the s_t terms on the left-hand side of the normal equations the quantities z and v are sine or cosine functions that are numerically less than one, or are simple powers of t; hence for large t the added term $a^{t+1} z_{-(t+1)} v_{-(t+1)}$ becomes smaller and smaller as t increases. It can in fact be shown that for large t all S terms on the left-hand side approach constant values. As these values do not depend on the data, they can be worked out as a set of numerical values $S_\infty(z,v)$ as a preliminary to the forecasting exercise.

Notice that this approach exactly parallels the methods discussed for fitting a linear trend model. In that case discounted least squares gave a set of normal equations which for large t led to the simplification of the recurrence relations of the double exponential smoothing method. If the values $S_\infty(\ )$ are used in the normal equations, from the start of the forecasting method, we obtain a method that approximates the discounted least squares solution in the same sense that the double exponential smoothing method does in the linear trend situation. Further, if explicit solutions for the fitted constants are obtained by solving the normal equations, then these solutions, though rather complicated, can be expressed in recurrence form. They can also, as has happened before, be expressed in error-correction form. Suppose we denote the one-stop-ahead error by e_{t+1}. Then

$$e_{t+1} = x_{t+1} - b_{2,t} \sin \frac{2\pi.1}{n} - b_{3,t} \cos \frac{2\pi.1}{n}$$

Then using the results of (b) we can show that

$$b_{2,t+1} = b_{2,t} \cos \frac{2\pi}{n} - b_{3,t} \sin \frac{2\pi}{n} + k_2 e_{t+1}$$

$$b_{3,t+1} = b_{2,t} \sin \frac{2\pi}{n} + b_{3,t} \cos \frac{2\pi}{n} + k_3 e_{t+1}$$

The parameters k_2 and k_3 controlling the amount of response are direct functions of a. They can be found using the normal equations, with t as origin, by replacing $b_{2,t}$ by k_2, $b_{2,t}$ by k_3 and $S_t(z,v)$ by the value of z at the present, zero time. However, as with Holt's method, we may regard these results not as an approximation to the discounted least squares method but as a new method in which the user can control the two parameters k_2 and k_3 to get the best results.

It should be noted that, for both methods, as the number of model parameters increases we need to make greater use of past data to get viable estimates. Thus the optimum value of a is likely to get closer to the global model value of one; the values of k_1 and k_2 are likely to be correspondingly small.

It is hoped that the above discussion of one example has been sufficient to give the reader enough understanding to construct the normal equations for other models and, possibly, to get these to a form that can provide a workable solution. The usual presentation of these results makes use of a matrix presentation that gives a very elegant way of studying the method. The reader is referred to two books by R. G. Brown (1963, 1967) that deal with this approach clearly and thoroughly.

8.4 Stochastic methods

In recent years there has been a growing interest in using stochastic models to forecast seasonal data. The basis of this approach is to construct models in which there is a stochastic relationship between each observation and (a) those close to it in time and also (b) observations in the same season of previous periods. To construct such models we make use of differences between observations in the same season of successive periods. Thus we would define

$$\nabla^r x_t = x_t - x_{t-r}$$

Quoting an example from Box and Jenkins (1970), we might, for each month separately, construct a model

$$\nabla^{12} x_t = \alpha_t - \theta \alpha_{t-12}$$

where the α values form a stochastic process. As the second stage we now build a model for the process of α values. We might, for example, have

$$\nabla \alpha_t = \epsilon_t - \beta \epsilon_{t-1}$$

where the ϵ_t values form an independent sequence. Substituting for α_t from this second model in the first and expanding, we finally have the model

$$x_t - x_{t-1} - x_{t-12} + x_{t-13} = \epsilon_t - \beta\epsilon_{t-1} - \theta\epsilon_{t-12} + \theta\beta\epsilon_{t-13}$$

This can now be used to obtain forecasts using the method discussed for stochastic models. This model provides a very elegant form of seasonal model using only two parameters to describe a fairly complex situation. A somewhat similar approach is used by Grether and Nerlove (1970), who use mixed autoregressive—moving average models independently for the trend—cycle and the seasonal components of a classical additive seasonal model. A detailed discussion of the methods for identifying and fitting such models is beyond the scope of this book, but is given in the book by Box and Jenkins (1970).

An alternative approach that leads to similar forecasting formulae to the above is based on using a regression of x_t on recent values and values a year or so ago. Thus the model might look like

$$x_t = b_0 + b_1 x_{t-1} + b_2 x_{t-12} + b_3 x_{t-13} + e_t$$

The actual terms used in the model can be determined by the usual methods for selecting variables. A brief discussion of this approach is given in section 15.4.

8.5 Comparison of methods

The previous sections have looked in some detail at the problems and methods of seasonal forecasting. The first motive for such an extensive treatment was the fact that seasonal variation occurs frequently in practice. The second was that it provided opportunity to discuss forecasting using models that contain several different components showing different types of behaviour. We have looked at three approaches to forecasting in seasonal situations. Which method is best in any situation is clearly a matter of practical investigation. There is some evidence to guide one, e.g. Reid (1972), Groff (1973). If the irregular, random, component is small, then the seasonal index methods are probably best. If this component is large, however, it makes the estimation of the seasonal components difficult and the Fourier methods are likely to be better. Fourier methods clearly work best when the seasonals form a fairly smooth pattern. Fourier methods have an advantage when we are dealing with, say, 52 weeks in a year or 168 hours in a week. The manipulation of more than 50 seasonal indices is unlikely to give better results than a Fourier model with, say, 10 parameters. If the seasonals form a very irregular pattern within the year, then the use of the separate seasonal index for each season becomes advisable. Where there is ample data and time for investigation, the identification and fitting of a suitable stochastic model may lead to good forecasts. The problem of identifying an appropriate

149

model is however far from easy, e.g. see Chatfield and Prothero (1973), Box and Jenkins (1970). Providing good forecasts may not, however, be the only consideration. The estimates of current trends, seasonal indexes and the deseasonalized data provided by the seasonal index methods and Fourier methods, but not directly by the stochastic methods, are often important supplementary information for decision making in seasonal situations.

References

Box, G. E. P. and Jenkins, G. M. (1970). *Time Series Analysis and Forecasting.* Butterworths, London.
Box, G. E. P., and Jenkins, G. M. (1973). Some comments on a paper by Chatfield and Prothero and on a review by Kendall. *J. Roy. Statist. Soc. A* 135, 337—345.
Brown, R. G. (1963). *Smoothing, Forecasting and Prediction of Discrete Time Series.* Prentice-Hall. Englewood Cliffs. N.J.
Brown, R. G. (1967). *Decision Rules for Inventory Management.* Holt Rinehart and Winston.
Chatfield, C. and Prothero, D. L. (1973a). Box—Jenkins seasonal forecasting. *J. Roy. Statist. Soc. A.* 136, 295—336.
Chatfield, C. and Prothero, D. L. (1973b). A reply to some comments by Box and Jenkins. *J. Roy. Statist. Soc. A.* 136, 345—352.
Dauten, C. A. (1961). *Business Cycles and Forecasting.* South-Western Publishing Co., Cincinnati, Ohio.
Davis, H. T. (1941). *The Analysis of Time Series.* Cowles Commission, Yale.
Durbin, J. and Murphy, M. J. (1975). Seasonal adjustment based on a mixed additive—multiplicative model. *J. Roy. Statist. Soc. A.* 138, 385—410.
Duvall, R. M. (1966). Time series analysis by modified least-squares techniques. *JASA*, 61, 152—165.
Grether, D. M. and Nerlove, M. (1970). Some properties of optimal seasonal adjustment. *Econometrica,* 38, 682—703.
Groff, G. K. (1973). Empirical comparison of models for short range forecasting. *Management Science,* 20, 22—30.
Harrison, P. J. (1965). Short-term sales forecasting. *J. Roy. Statist. Soc. C,* 14, 102—139.
Jorgenson, D. W. (1967). Seasonal adjustment of data for econometric analysis. *JASA*, 62, 137—140.
Kendall, M. G. (1973). *Time Series.* Griffin Co. Ltd., London.
Kendall, M. G. and Stuart, A. (1966). *The Advanced Theory of Statistics,* Vol. 3. Griffin, London.
Lovell, M. C. (1963). Seasonal adjustment of economic time series and multiple regression analysis. *JASA*, 58, 993—1010.
Nettheim, N. F. (1964). Fourier methods for evolving seasonals. *JASA*, 60, 492—502.
Reid, D. J. (1972). A comparison of forecasting techniques in economic time-series. In M. J. Branson and others (Eds.), *Forecasting in Action.* Op. Res. Soc. and Soc. for Long Range Planning, London.
Slutsky, E. (1937). The summation of random causes as the source of cyclic processes. *Econometrika, 5.*

Chapter 9

Growth curves

9.1 Models for growth curves

The aim of this section is to introduce some non-linear models that are finding increasing use in forecasting. These models are used in situations with an element of growth (or decay) in them. Examples of such curves are shown in Figure 9.1. The features of these curves, that cannot easily be modelled by the use of polynomials, are the regions of very rapid change and of maximum (or minimum) values to which some of the curves tend. Let us start by considering a number of situations where such curves might occur.

(a) Many items, such as consumption of electricity or numbers of books published, seem to increase at approximately a constant percentage rate, e.g. a 5 per cent increase each year. Thus if the first year's sales are 100 units, the second year would be $100 \times 105/100 = 105$ and the third year would be $105 \times 105/100 = 110.25$. The plot of this data extended is shown in Figure 9.2 on ordinary graph paper. Figure 9.2 also shows the line of the logarithms of the data. The series of original data curve upwards as sales expand. The logarithms of the data follow a straight line. The reason for this is that we could write the third year's sales as $100 \times (105/100)^2$, the fourth as $100 \times (105/100)^3$, etc.

Taking logarithms gives log sales as

$$\log 100 + 2 \log \frac{105}{100}$$

$$\log 100 + 3 \log \frac{105}{100}$$

etc. Thus the log sales increase by a constant amount $\log 105/100$

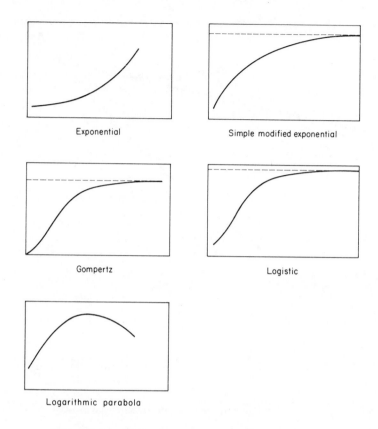

Figure 9.1   Sketches of some typical growth curves

each year. We could write the general model for such sales as

$$x_t = \alpha r^t$$

or

$$\log x_t = \log \alpha + t \log r$$

where in the above example $\alpha = 100$ and $r = 1.05$. It is often
convenient to write $r = e^\beta$ and use logarithms to the base e. The
above models thus become

$$x_t = \alpha e^{\beta t}$$

called the exponential growth curve, and

$$\log x_t = \log \alpha + \beta t$$

This model gives a good description of many sets of data and can
be very good at producing short-term forecasts. However, when we

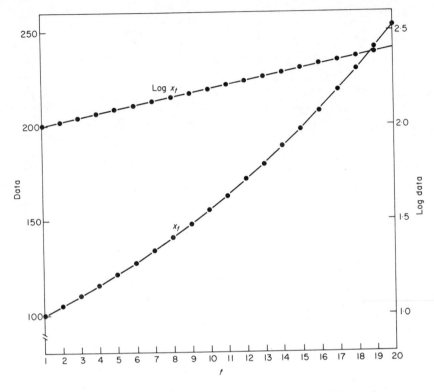

Figure 9.2   Plot of data and log data for exponential growth

consider longer term forecasts, there must always be an element of doubt that any real series will continue to increase with greater and greater rapidity, as is implied by the exponential curve. It is more likely that the rate of growth will die off as time goes on and, possibly, that some sort of saturation level will eventually be reached.

(b) Situations where the figures reach some fairly steady maximum or minimum value are common. The lines to which these curves tend, shown dotted in Figure 9.1, are called asymptotes, and these three curves are sometimes called asymptotic growth curves. It often happens that when a new product is launched, which competes with existing products in its field, it will expand into the field from zero sales until it slowly ceases expansion at some certain percentage of the total sales in the field. Thus a model for the sales of this product during the expansion phase as a percentage of the total market would most appropriately be one of the asymptotic growth curves.

(c) Occasionally we have an item, such as a stylized product, that will only sell for a certain period, e.g. a year in the case of fashion clothes. The monthly sales for such an item might follow a curve, such as the logarithmic parabola in Figure 9.1, or, alternatively, we might try to model the total sales up to any given month by one of the asymptotic growth curves.

The growth curves of Figure 9.1 thus provide a useful class of models. Lewandowski (1974) discusses a large number of growth curves of use in forecasting. Those mentioned in Figure 9.1 are the most common. It is not proposed to carry out a detailed mathematical study of these curves individually. In Table 9.1 the main formulae and properties associated with these curves are given. They are all described by models with at most three parameters. Though these parameters can take any values, their use in prediction tends to impose conditions on the parameters, as shown in the table. The condition shown without brackets has been assumed when calculating the values of x_t and the slope when t approaches minus and plus infinity. Table 9.1 illustrates the significance of some of the parameters in terms of the shape of the curve. Referring to Table 9.1 and Figure 9.1, it is clear that the Gompertz and logistic curves are very similar. The meanings of the various parameters are intuitively the same: α gives the position of the asymptote, γ controls the scale of the graph along the time axis and β the positioning of the point of inflexion.

The problem of deciding from a set of data which curve is appropriate, the identification problem, is particularly difficult in forecasting. One's initial data usually come from the left-hand end of the axis and at this end the curves look pretty much the same. There are two common ways of deciding which model is appropriate. The simplest is to plot the data on special graph paper using axes designed to give straight lines if the data come from the appropriate model.

A second approach to identifying the appropriate model is to make use of what are called 'slope characteristics'. If we calculate the slope and express it in terms of x_t and t, we obtain the expressions given under the heading 'slope equation' in Table 9.1. If these equations are rearranged so that the left-hand side takes the form given under the heading 'slope characteristic', then the right-hand side has the form of a linear trend, possibly with zero slope. For example, for the logarithmic parabola

$$\text{slope}/x = \beta + 2\,\gamma t$$

and for the Gompertz curve

$$\log\,(\text{slope}/x) = (\log\,\gamma\beta) + \gamma t$$

Thus if we plot the various slope characteristics against time, then we will get a straight line in the one case where the slope characteristic is

Table 9.1 Growth curves

Name	Formula $x_t =$	Usual condition	$t = -\infty$		$t = 0$		$t \to \infty$		Point of inflexion (x_p, t_p)	Slope equation	Slope characteristic
			x slope		x	slope	x	slope			
Exponential	$\alpha e^{\beta t}$	$\beta > 0$ $(\alpha > 0)$	0	0	α	$\alpha\beta$	∞	∞	None	$\text{slope}_t = \beta x_t$	slope/x
Simple modified exponential	$\delta + \alpha e^{\beta t}$	$\beta < 0$ $(\alpha < 0)$	$-\infty$	∞	$\delta + \alpha$	$\alpha\beta$	δ	0	None	$\text{slope}_t = \beta(x_t - \delta)$	**log** slope
Logarithmic parabola	$\alpha e^{\beta t + \gamma t^2}$	$\delta > 0$ or $\gamma < 0$ $(\alpha > 0)$	∞	0	α	$\alpha\beta$	∞	∞	Two	$\text{slope}_t = (\beta + 2\gamma t)x_t$	slope/x
Gompertz curve	$\alpha e^{(\beta e^{\gamma t})}$	$\gamma < 0$ $(\alpha > 0)$ $(\beta < 0)$	0	0	αe^{β}	$\alpha\beta\gamma e^{\beta}$	α	0	$1 + \beta e^{\gamma t_p} = 0$ $x_p = \alpha/e$	$\text{slope}_t = \gamma\beta e^{\gamma t} x_t$	log(slope/x)
Logistic curve	$\alpha/(1 + \beta e^{\gamma t})$	$\gamma < 0$ $(\alpha > 0)$ $(\beta > 0)$	0	0	$\dfrac{\alpha}{1 + \beta}$	$\dfrac{-\alpha\gamma\beta}{(1 + \beta)^2}$	α	0	$1 - \beta e^{\gamma t_p} = 0$ $x_p = \alpha/2$	$\text{slope}_t = -\dfrac{\gamma\beta}{\alpha} e^{\gamma t} x_t^2$	$\sim$ log(slope/x^2)

the right one for the model. A detailed study of the use of slope characteristics is given by Gregg, Hassell and Richardson (1964).

The method has two practical difficulties whose importance depends very much on the smoothness of the data available. There is first the problem of measuring the slope at different times. This can be done approximately by taking first differences and smoothing these with a moving average. Alternatively, one can smooth first and then take differences or find the slope of a fitted trend line. Neither approach is very satisfactory. The second problem comes from the fact that the method depends on an eye comparison of different plots to see which looks most like a straight line. One can, however, be misled in this, since the vertical scales are all in different units. Within the forecasting context it is best to follow a preliminary study of the data, using one of the above two methods, by a study of the quality of forecasts that the most appropriate models produce.

An important property of the growth curves of Table 9.1 is that by suitable choice of transformation they can be transformed either to a linear model or to a simple modified exponential. The nature of these transformations is displayed in Table 9.2. As long as we are investigating deterministic models, these transformations can be quite useful in helping us to inspect the data to decide what model is appropriate; we could also use the linear form as the basis of extrapolation. The same is also true if there is a very small random component in the model. If, however, there is a large random element superimposed on a

Table 9.2 Transformations of growth curves

Growth curve	Linear form
Exponential	$\log x_t = \log \alpha + \beta t$
Simple modified exponential	$\log (x_t - \delta) = \log \alpha + \beta t$
Logarithmic parabola	$\log x_t = \log \alpha + \beta t + \gamma t^2$
	Simple modified exponential form (and linear form)
Gompertz curve	$\log x_t = \log \alpha + \beta e^{\gamma t}$ and hence $\log (\log x_t - \log \alpha) = \log \beta + \gamma t$
Logistic curve	$\dfrac{1}{x} = \dfrac{1}{\alpha} + \dfrac{\beta}{\alpha} e^{\gamma t}$ and hence $\log \left(\dfrac{1}{x_t} - \dfrac{1}{\alpha} \right) = \log \dfrac{\beta}{\alpha} + \gamma t$

Note that log here is $\log_e$, the natural logarithm.

growth curve, the use of transformations to linear form, followed by estimation of parameters and extrapolation, may lead to poor forecasts.

9.2 Fitting growth curves

There is a large literature dealing with the fitting of growth curves (e.g. Harrison and Pearce, 1972; Lewandowski, 1974). Rather than attempt the rather daunting task of giving a survey of the subject, the aim of this section is to indicate by examples a number of approaches that can be adopted.

9.2.1 Transformation, fitting and retransforming

This approach is very common and is based on the transformations shown in Table 9.2. For example, a set of data from an exponential situation would be transformed by taking logs. A least squares or discounted least squares line is then fitted as described in Chapter 5 and antilogs are then taken to give the fitted exponential growth curve. The fitting of the logged data assumes that the model being used is

$$\log_e x_t = \log_e \alpha + \beta t + \epsilon_t, \ \mathrm{Var}(\epsilon_t) = \sigma^2$$

i.e.

$$x_t = \alpha e^{\beta t + \epsilon_t}$$

Now we normally regard the form $\alpha e^{\beta t}$ as giving the expected curve with random deviations from it. Unfortunately, the expectation of x_t in the above model is $\alpha e^{\beta t + \sigma^2/2}$ and not $\alpha e^{\beta t}$, as is implicitly assumed by the above method. Thus even if we obtained a perfect fit and forecast x_{t+h} by

$$\hat{x}_{t+h} = \alpha e^{\beta(t+h)}$$

we would have a bias given by

$$E(e_{t+h}) = E(x_{t+h} - \hat{x}_{t+h}) = \alpha e^{\beta(t+h)}(e^{\sigma^2/2} - 1)$$

This bias increases with σ^2 and also increases as one climbs up the growth curve. We will discuss in Chapter 16 methods to adjust for systematic bias in forecasts. For the moment it is sufficient to draw attention to the problem of the possible introduction of bias when transformations are used.

9.2.2 Least squares and discounted least squares

Consider as a model for exponential growth the expression

$$x_t = \alpha e^{\beta t} + \epsilon_t$$

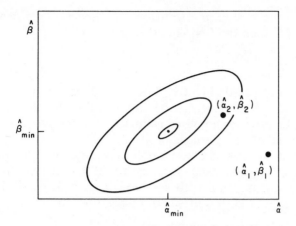

Figure 9.3   Contours of S

The method of least squares requires us to choose estimators $\hat{\alpha}$ and $\hat{\beta}$ to minimize

$$S = \sum_t \hat{\epsilon}_t^2 = \sum_t (x_t - \hat{\alpha}e^{\hat{\beta}t})^2$$

We can do this by a procedure that simply works on the assumption that S is a surface in the $(\hat{\alpha},\hat{\beta})$ plane, such as that shown by the contours in Figure 9.3. Various pairs of values $(\hat{\alpha}_1,\hat{\beta}_1)$, $(\hat{\alpha}_2,\hat{\beta}_2)$, ... are substituted and the values $S_1, S_2, ...$ calculated until the minimum point S_{min} is found. The corresponding values $(\hat{\alpha}_{min}, \hat{\beta}_{min})$ are then the least squares estimators. Exactly the same procedure may be adopted if we use a discounted least squares criterion.

There are two classes of method for choosing the sequence of points $(\hat{\alpha}_1,\hat{\beta}_1)$, $(\hat{\alpha}_2,\hat{\beta}_2)$ in the search for the minimum. In the first class, methods are devised for moving towards the minimum by obtaining S at a sequence of points and using these values to guide us to where to move to next. Such methods are often termed 'hill climbing' methods. The second class of method makes specific use of the form of the function, the growth curve. Thus the slope of the curve being fitted can be used to choose a suitable sequence of points. Draper and Smith (1966) discuss three such methods. The method of Marquardt (1963) and its various developments are particularly relevant to fitting growth curves (e.g. see Barham and Drane, 1972).

If we try to find $(\hat{\alpha}_{min}, \hat{\beta}_{min})$ theoretically by using calculus, we obtain the following pair of normal equations:

$$\sum_t x_t e^{\hat{\beta}t} - \hat{\alpha} \sum_t e^{2\hat{\beta}t} = 0$$

$$\sum_t tx_t e^{\hat{\beta}t} - \hat{\alpha} \sum_t te^{2\hat{\beta}t} = 0$$

158

which do not have a simple solution. If, however, we can obtain an estimate $\hat{\beta}_1$ of β, then the first equation gives

$$\hat{\alpha}_1 = \sum_t x_t e^{\hat{\beta}_1 t} / \sum_t e^{2\,\hat{\beta}_1 t}$$

which can be evaluated directly. The second equation can be used to give a check on the closeness of the solution to the least squares solution and to produce a revised estimate of β; the process can then be repeated until satisfactory solutions are found.

In both least squares and discounted least squares it is assumed that the residuals $\hat{e}_i$ have the same variance for all t. It will often occur that the variablity will increase with the level of growth curve. In this case the methods should be modified to replace the term $\hat{e}_i^2$ by $w_i\hat{e}_i^2$, where

$$w_i = 1/(\text{estimated variance of } x_i)$$

The intuitive logic of this is that if the variance is large less notice should be taken of a large $\hat{e}_i$. One way of obtaining such an estimated variance is to model the variance as a function of level of the curve. For example, one method uses

Variance is proportional to (level)^k

where the level of the curve is found using a weighted moving average of the data and k is some power determined to give the best fit (see Harrison and Pearce, 1972).

9.2.3 A data sectioning method

An approach that can be applied to several of the growth curves to obtain simple approximate fits is to divide the data into sections and work out some statistics for each section. The unknown parameters are then found from these statistics. To illustrate this approach consider the modified exponential

$$x_t = \delta + \alpha e^{\beta t} + \epsilon_t$$

The normal equations for this can be solved simply for $\hat{\delta}$ and $\hat{\alpha}$, but the $e^{\hat{\beta}t}$ term is not separable. If, as before, we can get an estimate of β, then the normal equations can be used to provide estimates of δ and α. Suppose we have $3n$ observations and we consider the model values

$$x_1 = \delta + \alpha e^{\beta} + \epsilon_1$$
$$x_2 = \delta + \alpha e^{2\beta} + \epsilon_2$$
$$x_n = \delta + \alpha e^{n\beta} + \epsilon_n$$

$$\overline{x}_1 = \delta + \alpha \overline{e^{i\beta}} + \overline{\epsilon_1}$$

where $\bar{x}_1$ is the average of the first n observations, $\bar{\epsilon}_1$ of the first n ϵ values and $e^{\overline{i\beta}}$ is the average of e^β, $e^{2\beta}$, . . . , $e^{n\beta}$. If we do the same for the next n observations, we get

$$\bar{x}_2 = \delta + \alpha e^{n\beta} e^{\overline{i\beta}} + \bar{\epsilon}_2$$

since the terms in $e^{t\beta}$ all have the form $e^{(n+i)\beta}$. Similarly, for the last n observations

$$\bar{x}_3 = \delta + \alpha e^{2n\beta} e^{\overline{i\beta}} + \bar{\epsilon}_3$$

Replacing the $\bar{\epsilon}$ by their expected values of zero, we can eliminate δ by using

$$\bar{x}_3 - \bar{x}_2 \stackrel{\frown}{=} \alpha e^{\overline{i\beta}} e^{n\beta} (e^{n\beta} - 1) \quad (\stackrel{\frown}{=} \text{means approximately equal to})$$

and

$$\bar{x}_2 - \bar{x}_1 \stackrel{\frown}{=} \alpha e^{\overline{i\beta}} (e^{n\beta} - 1)$$

so we have as an estimator

$$e^{n\hat{\beta}} = (\bar{x}_3 - \bar{x}_2)/(\bar{x}_2 - \bar{x}_1)$$

This can be used with the equations above or the normal equations to obtain estimators of δ and α. Gregg, Hassell and Richardson (1964) give details of these methods and tables to simplify the calculations.

9.2.4 Other ways of producing a linear form

The form of the slope equation can suggest other ways of obtaining linear equations. For example, in the case of the logistic curve we can remove the $e^{\gamma t}$ term by using the basic equation of the model and obtain

$$\frac{1}{x_t} \text{ slope}_t = \gamma - \frac{\gamma}{\alpha} x_t$$

which we can write as

$$z_t = \gamma - \theta x_t$$

which is an ordinary linear model. The values of z_t can be found, roughly, from the data using the methods suggested for finding slope characteristics. We can then apply ordinary or discounted least squares to estimate γ and θ; $\hat{\alpha}$ then comes from $-\hat{\gamma}/\hat{\theta}$. To estimate β we note from the model, putting ϵ_t as its expected value of zero, that

$$\beta \stackrel{\frown}{=} e^{-\gamma t} \left(\frac{\alpha}{x_t} - 1 \right)$$

So if we define

$$u_i = e^{-\hat{\gamma} i} \left(\frac{\alpha}{x_i} - 1 \right)$$

an estimate of β is then given by the average u_i thus:

$$\hat{\beta} = \sum_{i=1}^{t} u_i / t$$

9.3 Forecasting growth curves

Much of the literature on forecasting makes a distinction between long-term and short-term forecasting. From the statistical viewpoint this is not a meaningful distinction, as all our forecasting is based on model-fitting and extrapolation. Whether we extrapolate for a lead time of 1 or of 100, the basic procedure is the same. Ideally we choose the best model for both lead times. However, if we leave statistical considerations aside, then it is clear that, for a lead time of 1, several different models may well give similar forecasts and a choice of model based on simplicity of use might well be appropriate. Conversely, for a long-term forecast, a model that correctly follows the curvature and possible asymptotic properties of the data will justify the extra effort required to obtain a good fit. A consequence of these considerations is that growth curves have mainly been used for long-term forecasting. For such a use considerable care must be given to the consideration of (a) the structure and (b) the stability of the situation.

(a) For the structure we need to identify not only the type of growth curve to use but also the way the random variation is involved in the structure. For example, an exponential growth might be of the form

$$x_t = \alpha e^{\beta t} + \epsilon_t$$

or

$$x_t = \alpha e^{(\beta t + \epsilon_t)}$$

In the former case least squares methods should be applied to the data directly; in the latter case to $\log_e x_t$. In this latter case the model for $\log x_t$ is the ordinary linear trend model of Chapter 5. Having obtained estimates α and β the natural forecast will be

$$\bar{x}_{t+h} = \alpha e^{\hat{\beta}(t+h)}$$

in both cases. As indicated before, the non-linear form of the model means that the forecasts $\hat{x}_{t+h}$ will in general be biased, the amount of bias increasing with σ_ϵ^2. There are no clear-cut ways of identifying and using the 'right' structure. If adequate historical data is available, several alternatives should be considered and the forecast errors analysed using the methods discussed in Chapter 15.

(b) In studying the likely stability of the model a thorough study is needed of the nature of the variable being forecast. For a growth curve to be assumed stable over a long period, the growth shown must be a natural result of an existing and developing structure of, for example, market forces. For long-term forecasting the assumption of stability needs to be related to the total environment in which the model operates. Though a study is beyond the scope of this book, it is worth noting that the types of growth curve in common use can all be derived from more general models of some situations. These models describe the ways in which the variable x_t relates to various forces that act upon it. For example, suppose x_t are the weekly sales in a market with ultimate potential weekly sales of S. A possible model would assume that the rate at which x_t will increase will depend on the number of potential customers, who do not yet buy the product. Thus x_t will increase most rapidly at first but slow down as S is approached. The model is thus

Rate of change of $x_t = \beta(S - x_t)$

where β is a constant of proportionality. A mathematical analysis of this shows that x_t must follow the growth curve.

$$x_t = S(1 - e^{-\beta t})$$

which is the special case of the simple modified exponential curve illustrated in Figure 9.1.

References

Barham, R. H., and Drane, W. (1972). An algorithm for least squares estimation of non-linear parameters when some of the parameters are linear. *Technometrics*, 14, 757—766.

Draper, N. R., and Smith, H. (1966). *Applied Regression Analysis*. John Wiley and Sons, New York.

Gregg, J. V., Hassell, C. H., and Richardson, J. T. (1964). *Mathematical Trend Curves; An Aid to Forecasting*. ICI Monograph. Oliver and Boyd. Edinburgh.

Harrison, P. J., and Pearce, S. F. (1972). The use of trend curves as an aid to market forecasting. *Ind. Mark. Manage.*, 2, 149—170.

Lewandowski, R. (1974). *Prognose und Informations Systeme*. W. de Gruyter, Berlin.

Marquardt, D. W. (1963). An algorithm for least squares estimation of non-linear parameters. *J. Soc. Ind. Appl. in Math.*, 11, 431.

Chapter 10

Probabilistic models

10.1 Introduction

In previous chapters interest has centred on forecasting some variable x_t. In a number of situations the item of central interest is an event or a number of possible events. The model used then concentrates on the probabilities of these events occurring. The forecasting problem in such situations is that of estimating the future values of these probabilities. For example, the prices of certain foodstuffs tend to be either stable, with the basic price remaining constant from day to day, or unstable, with price changes taking place. In obtaining price forecasts we are faced with forecasting the probabilities of the two alternative events, 'prices stable' or 'prices unstable', at a future time. Another situation which faces us more directly with probability considerations is where we are forecasting a discrete variable that may only take on a limited range of values. For example one might require to forecast, in a changing situation, the number s_t of vehicles out of a small fleet of five that would be in use at a future time. Such a problem requires a 'forecast' of the future probabilities of demands occurring for use of a vehicle. The aim of this chapter is to discuss briefly some approaches to the forecasting of probabilities.

10.2 Forecasting probabilities

We now consider the problem of how to estimate probabilities in situations where there may be slow changes in these probabilities. We are dealing with a situation, analogous to that discussed in Chapter 4, where there are simple and well-known methods for finding estimates of probabilities where these are globally constant, but where we must now regard the model as only locally constant.

In previous chapters we have frequently made use of 'discounted least squares' to fit a model locally for forecasting use. In this approach the usual least squares criterion of fit

$$\sum_{i=1}^{t} (\text{residual})_i^2$$

was replaced by

$$\sum_{r=0}^{t-1} w_r (\text{residual})_{t-r}^2$$

where the residual was the error between the observed value and the fitted value and the weights w_r were chosen to give more emphasis to recent data than past data. In fitting probability models the method of least squares is often replaced by a method called the method of maximum likelihood. The basis of this method is first to write down the probability of the events that were observed. This probability is a function of the model parameters and is called the likelihood. Next we choose the model parameters so that the events observed had a higher probability of occurrence than events that were not observed. Thus if the probability of observing the discrete variable s_t at time t is $P_t(s_t;\theta)$, where θ is a parameter, the likelihood, probability, of observing data $s_1 \ldots s_t$ is

$$\mathcal{L}(\theta) = P_1(s_1;\theta)P_2(s_2;\theta) \ldots P_t(s_t;\theta)$$

We now choose the parameter θ to maximize this probability. Another way of expressing this is to take logarithms of both sides giving

$$L(\theta) = \log \mathcal{L}(\theta) = \sum_{i=1}^{t} \log P_i(s_i; \theta)$$

The value of θ which maximizes this also maximizes $\mathcal{L}(\theta)$. Thus the estimator $\hat{\theta}$ can be chosen by maximizing $\mathcal{L}(\theta)$ or $L(\theta)$, whichever is most convenient.

A natural way to modify this global method of estimation, to deal with local changes in parameter values, is to introduce a discounting factor w_r, as above. The discounted versions of $\mathcal{L}(\theta)$ and $L(\theta)$, denoted by $\mathcal{L}_w(\theta)$ and $L_w(\theta)$ and called the discounted likelihood and discounted log likelihood, are

$$\mathcal{L}_w(\theta) = \{P_t(s_t;\theta)\}^{w_0} \{P_{t-1}(s_{t-1};\theta)\}^{w_1} \ldots \{P_1(s_1;\theta)\}^{w_{t-1}}$$

and

$$L_w(\theta) = \sum_{r=0}^{t-1} w_r \log P_{t-r}(s_{t-r}; \theta)$$

It will be seen that the discounting is applied to the term $\log P_{t-r}(s_{t-r};\theta)$ in exactly the same way as it is applied to $(\text{residual})_{t-r}^2$ in discounted least squares.

Let us illustrate this method by considering the food price example of section 10.1. Let p be the probability of the price being stable (S) at any time and, hence, $1-p$ be the probability of it being unstable (U). Thus $P_{t-r}(S_{t-r};\theta)$. in the discounted likelihood, is just p when S occurs and $1-p$ when U occurs. If we use our simple exponential discounting weights, $w_r = a^r$, the terms in the discounted log likelihood are as shown for a particular set of data in Table 10.1(a). We can obtain a local estimate of p, $\hat{p}$, by finding the value of p that makes the quantity

Table 10.1 Local estimation of a probability

(a) Theory

Time	Stable S or unstable U	Probability	Contribution to $L_w(p)$
t	S	p	$\log p$
$t-1$	S	p	$a \log p$
$t-2$	S	p	$a^2 \log p$
$t-3$	U	$1-p$	$a^3 \log(1-p)$
$t-4$	S	p	$a^4 \log p$
$t-5$	U	$1-p$	$a^5 \log(1-p)$
$t-6$	U	$1-p$	$a^6 \log(1-p)$
$t-7$	S	p	$a^7 \log p$
$t-8$	U	$1-p$	$a^8 \log(1-p)$
$t-9$	U	$1-p$	$a^9 \log(1-p)$
		Total	$L_w(p)$

$\hat{p}_t$ maximizes $L_w(p)$

(b) Calculation (not using a recurrence form)

Time	S or U	δ_{t-r}	a^r	$a^r \delta_{t-r}$	a^r	$a^r \delta_{t-r}$	a^r	$a^r \delta_{t-r}$
					$a=1$		$a=0.9$	
t	S	1	1	1	1	1	1.000	1.000
$t-1$	S	1	a	a	1	1	0.900	0.900
$t-2$	S	1	a^2	a^2	1	1	0.810	0.810
$t-3$	U	0	a^3	0	1	0	0.729	0.000
$t-4$	S	1	a^4	a^4	1	1	0.656	0.656
$t-5$	U	0	a^5	0	1	0	0.590	0.000
$t-6$	U	0	a^6	0	1	0	0.531	0.000
$t-7$	S	1	a^7	a^7	1	1	0.478	0.478
$t-8$	U	0	a^8	0	1	0	0.430	0.000
$t-9$	U	0	a^9	0	1	0	0.387	0.000
	Totals		A	D	10	5	6.511	3.844

$\hat{p}_t = D/A$

$\hat{p}_t = 5/10$
$= 0.5$

$\hat{p}_t = \dfrac{3.844}{6.511} = 0.590$

$L_w(p)$ as large as possible. This looks a little complex, but we can simplify the formulation of the problem by introducing a quantity called an indicator variable δ_{t-r}. This is set at one or zero according to whether S or U occurs at time $t-r$, as is seen in Table 10.1(b). The contributions to $L_w(p)$ can now be written as

$$a^r[\delta_{t-r}\log p + (1-\delta_{t-r})\log(1-p)]$$

Maximizing $L_w(p)$ with respect to p can now be shown to lead to

$$\hat{p}_t = \sum_{r=0}^{t-1} a^r \delta_{t-r} \Big/ \sum_{r=0}^{t-1} a^r = D/A$$

Thus $\hat{p}_t$ is the discounted average of the indicator variables. Table 10.1(b) shows the fairly simple layout required to calculate $\hat{p}_t$ at a fixed point in time. As usual D and A can be simply updated as time goes by. If $a = 1$, then $\hat{p}_t$ is the ordinary average of the indicator variables, i.e. the proportion of occurrences of the stable state. In Table 10.1(b) for $a = 1$, we have $\hat{p} = 5/10 = 0.5$, as for half the weeks the state is stable. If $a = 0.9$, this becomes 0.59, indicating the preponderance of stable states in recent weeks. The above expression for $\hat{p}$, though derived for the case of two possible events, is in fact quite general. We obtain the local estimate of the probability of any event, of a number of possible events, from this formula by using $\delta = 1$ when the event of interest occurs and $\delta = 0$ when any of the other events occur.

When t is large we can rewrite the expression for $\hat{p}$ approximately as

$$\hat{p}_t = (1-a)\sum_{r=0}^{t-1} a^r \delta_{t-r}$$

or, in recurrence form,

$$\hat{p}_t = (1-a)\delta_t + a\hat{p}_{t-1}$$

This form was suggested by Brown (1963) on rather different grounds.

Experience with using these methods suggests that in most situations it is advisable to take a value of a fairly close to one; otherwise the estimates $\hat{p}$ tend to fluctuate too much at random.

A second example of the use of this approach is given in Figure 10.1. This shows the use of the numerator, $\sum a^r \delta_{t-r}$, of $\hat{p}$ for the occurrence of the discrete variables $S = 1, 2, 3$ or 4 over a period of time. For $a = 1$ this is just the frequency of occurrence of the particular values of S, which are normally plotted as a bar chart. For the value $a = 0.99$ used it will be seen that the plots of the discounted frequencies show up a changing pattern in the frequency distribution. A variation on this plot can be used to observe any changes in the proportion of large errors occurring in an ordinary forecasting system.

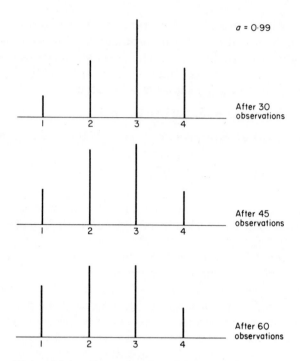

Figure 10.1   Example of a discounted frequency bar chart for a changing situation

10.3 State models

In this section we will briefly look at some forecasting situations where the interest is in the probability of the situation changing from one state to another. For example, in the price forecasting problem described in the last section it is found that in practice the probability of being in the stable state S next week depends on whether we are in the stable state S or the unstable state U this week. We are thus interested in the occurrence of transitions from one state to another. Table 10.2(a) shows the number of transitions of the four possible types for a particular set of vegetable prices. It will be seen that if we are in state S this week we are twice as likely to be in state S next week as in state U. In probability terms the 'transition' probability of going from state S to state S is about 2/3. Table 10.2(b) gives the notation for the possible transition probabilities and Figure 10.2 gives a diagramatic representation of such a situation.

A situation like this which takes the form of a 'chain of events' with a limited number of possible states is called a Markov chain. The theory and properties of Markov chains have been used extensively in forecasting, particularly within the market research area of application.

Table 10.2 State models

(a) Number of transitions of state

| | | State in next week | |
		S	U
State in	S	35	18
first week	U	11	36

(b) Notation for probabilities of transition

| | | State in next week | |
		S	U
State in	S	p_1	$1-p_1$
first week	U	$1-p_2$	p_2

(c) Global estimates of the transition probabilities from (a)

| | | State in next week | |
		S	U
State in	S	0.66	0.34
first week	U	0.23	0.77

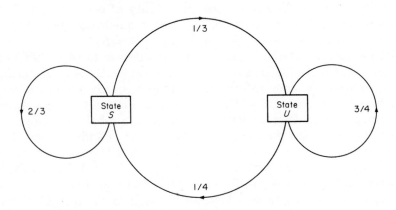

Figure 10.2   Transition diagram

One problem here lies in the estimating of the transition probabilities. It can be done globally as above by simply counting observed transitions. Thus of the 53 transitions from state S, 35 are to S, giving an estimate $\hat{p}_1 = 0.66$. Table 10.2(c) gives the four estimates. In the price forecasting situation these transition probabilities vary during the

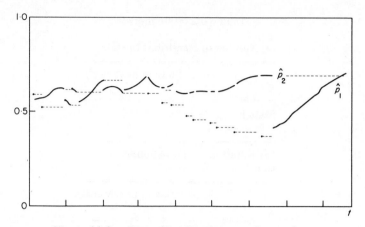

Figure 10.3   Plot of local estimates of p_1 and p_2

various seasons of the year and so p_1 and p_2 are best estimated locally using the discounted method of the previous section. Figure 10.3 shows the local estimate of p_1 and p_2 over a period of time for some price data. Notice that we only get information about p_1 when a transition is made from state S to S or U; thus only one of the curves varies at any one time. To illustrate how to use such transition probabilities, let us suppose that we know that we are in the stable state this week. We will be in state S next week with estimated probability $\hat{p}_1$ and state U with probability $1 - \hat{p}_1$. The estimated probability of being in state S two weeks hence is that of going through the sequence of states S,S,S or through S,U,S, which is $\hat{p}_1 \times \hat{p}_1 + (1 - \hat{p}_1) \times (1 - \hat{p}_2)$.

Before leaving Markov chains we consider a further forecasting problem that is similar in structure to the above. By way of example, Table 10.3 shows some information about a steady market in which four brands are competing. Part (a) gives the current market share of the four products. Part (b) gives the transition probabilities of product buyers between the brands. Thus buyers of brands A and B are completely loyal, but a person who buys brands D in year t has only a probability of 0.48 of buying it in year $t + 1$. Part (c) shows how this information can be used; to forecast the brand shares in year $t + 1$. We assume that because brand A held 50 per cent of the market in year t and had totally loyal customers it will hold an initial 50 per cent of the market. To this we must add the fraction 0.02 of the 10 per cent of the market that used to buy C, but which switches to A, and the fraction 0.22 of D brand's 10 per cent of the market, giving a total of 52.4 per cent. A basic problem here is that of estimating the probabilities of the transitions. One way of doing this, discussed by Chorofas (1965), is to carry out the reverse of the above calculations over a period of years for which market shares are known in successive years.

Table 10.3 A market share example

(a) Market share year t (%)

	Product			
	A	B	C	D
	50	30	10	10

(b) Transition probabilities

	To				
From	A	B	C	D	Total
A	1.0	0	0	0	1.0
B	0	1.0	0	0	1.0
C	0.02	0.03	0.95	0	1.0
D	0.22	0.30	0	0.48	1.0

(c) Contributions to market share year $t + 1$

	Year t shares (%)	Contributions to			
		A	B	C	D
A	50	50	0	0	0
B	30	0	30	0	0
C	10	0.20	0.30	9.50	0
D	10	2.20	3.00	0	4.8
Total year $t + 1$ market shares (%)		52.4	33.3	9.5	4.8

In the price forecasting example one state corresponded to the stable, constant price situation, S, which is modelled by

$$x_t = \mu$$

The other corresponded to the unstable price situation, U which in fact was modelled by

$$x_t = x_{t-1} + \epsilon_t$$

This example introduces the concept of using different possible states to describe situations in which changes of structure occur. A major application of this concept is to models such as that for the wandering mean introduced in section 7.6, i.e.

$$x_t = \mu_t + \epsilon_t$$
$$\mu_t = \mu_{t-1} + \gamma_t$$

In the stable state S, the ϵ_t and γ_t values come from distributions with zero mean and variances σ_ϵ^2 and σ_γ^2, respectively. If a very large value of ϵ_t occurs, then this shows as a transient in x_t. If a large value of γ_t occurs, μ_t will jump to a new level. Future means will start to wander from this new level; thus a major step change will have occurred in the mean level. The occurrence of such large values of ϵ_t and γ_t can be modelled by assuming the existence of two further states, U_ϵ and U_γ. In one, U_ϵ, the variance of ϵ_t is substantially larger than σ_ϵ^2, thus allowing the possibility of the transient, large ϵ_t. Similarly the other, U_γ, gives a large variance to γ_t, thus modelling the step-change. Clearly the system will only rarely be in states U_ϵ and U_γ. The probabilities of the transitions $S \rightarrow U_\epsilon$ and $S \rightarrow U_\gamma$ will thus be very small, and those for $U_\epsilon \rightarrow S$ and $U_\gamma \rightarrow S$ correspondingly high. Harrison and Stevens, using Bayesian methods, investigate a particular model of this type, which includes also a state for trend changes, U_δ. In their examples the variances involved in U_ϵ and U_γ are of the order of 100 times larger than σ_ϵ^2 and σ_γ^2, thus the occurrence of states U_ϵ or U_γ represent significant events in the progress of the data.

The value of using the state approach here is not that the forecasts make some allowance for future changes in state. This allowance must inevitably be small since the changes are relatively rare. The real value is that the model, as it is fitted to the data, detects past changes in state. This automatically improves the quality of the forecasts. Thus if a transient is detected, it will be treated as a transient and future forecasts will largely ignore it. Many other forecasting approaches will treat the transient as a respectable observation and it will lead to biased forecasts. Similarly, if a step-change is detected, the new forecasts made will be based on the levels of μ_t after the step and will tend to ignore data from before it. Other forecasting approaches will often take some time for the estimated level to move over to the region of the new level. The model incorporating the possibilities of transient and step-change states thus deals more effectively with data in which such features can occur. For technical details the reader is referred to the reference already given.

10.4 Forecasting criteria

In this chapter we have been examining the forecasting of one or more probabilities $p_1, \ldots, p_k$ using some forecasts $\hat{p}_1, \ldots, \hat{p}_k$. If our forecasts are sensible they will satisfy $0 \leqslant \hat{p}_i \leqslant 1$ and $\sum_{i=1}^{k} \hat{p}_i = 1$. It will usually occur that when we have obtained our observations we will only know whether certain things have happened or not; we will not know p_i. The consequence of these facts is that our usual mean square error criteria of a good forecast has no direct application. It is thus necessary to develop some new criteria upon which to judge forecasts. Let us start by considering the type of situation where p_i represents a proportion that will in fact be observed. In this case we can compare the forecasts $\hat{p}_i$ with the observed p_i. To derive such a comparison we use a method, given by Theil (1965), based on information theory. Suppose the event E_i has probability p_i and the statement is made 'E_i has happened'. If $p_i = 1$ this statement does not really tell you much: we already knew it was almost certain to happen. If, however, p_i is small, the statement is much more informative. The amount of information I_i in the statement is clearly an increasing function of $1/p_i$ and in fact we use

$$I_i = \log 1/p_i$$

as a measure of information. If we have k possibilities, the expectation of the amount of information to be obtained from our observations is

$$I = \sum_{i=1}^{k} p_i \log 1/p_i$$

$$= - \sum_{i=1}^{k} p_i \log p_i$$

Similarly, the expectation of the information in our forecasts $\hat{p}_i$ is

$$\hat{I} = - \sum_{i=1}^{k} p_i \log \hat{p}_i$$

If we had perfect forecasts these would be identical. The difference in information content of $\hat{p}$ and p is thus a measure of forecast accuracy. This is given by

$$\hat{D} = \hat{I} - I = \sum_{i=1}^{k} p_i \log \frac{p_i}{\hat{p}_i}$$

The smaller this quantity the better the forecasts, perfect forecasts giving $\hat{D} = 0$. Similarly, two different forecasts $\hat{p}_i$ and $\tilde{p}_i$ can be compared using $\hat{D} - \tilde{D}$, i.e.

$$\hat{I} - \tilde{I} = \sum_{i=1}^{k} p_i \log \frac{\tilde{p}_i}{\hat{p}_i}$$

If this is positive, $\tilde{p}$ will be judged the better forecast and conversely if it is negative. If we have no knowledge of p_i, but only the information that event E_j had occurred, we could assign $p_j = 1$ and all other $p_i = 0$. This would be equivalent to judging that method as best which assigned the highest probability to E_j. If we have several sets of observations, we would simply let p_i be the observed relative frequency of occurrence of E_i $(i = 1, \ldots, k)$.

Suppose that we have a sequence of observations at time $1, 2, \ldots, t$, with respective probabilities $p_1, p_2, \ldots, p_t$. The information content would be

$$- \sum_{r=1}^{t} \log p_i$$

If we were concerned with choosing parameters in p for forecasting purposes, as in section 10.2, we would wish to emphasize the most recent information. We therefore define the discounted information with weights w_r to be

$$- \sum_{r=0}^{t-1} w_r \log p_{t-r}$$

It has already been noted that if we know the future and could assign $p_i = 1$ to the appropriate future event E_i, then the information content of this perfect forecast would be zero. Thus we are aiming to produce forecasts with information at a minimum. This suggests that we choose the parameters in the formulae for the probabilities to minimize the discounted information. Referring back to section 10.2, this is just the same as maximizing the log discounted likelihood which is

$$\sum_{r=0}^{t-1} w_r \log p_{t-r}$$

This was the method used to obtain our forecasts in that section. Thus our brief discussion of information has given further support for the methods of obtaining probability forecasts in section 10.2.

10.5 Probability forecasting and decision making

The above discussion assumed that our interest centred in the probabilities themselves and that we had no further information about the practical situation. The usual object of calculating forecast probabilities $\hat{p}_i$ is to enable a decision d to be made. Let d_i be the decision to be made if one was certain that the event E_i would occur. Let C_{ij} be the cost involved if decision d_i is made and event E_j actually occurs. The expected cost of decision d_i is

$$\sum_{j=1}^{k} C_{ij} p_j$$

p_j being the true probability of E_j occurring. As we do not know this, we must use $\hat{p}_j$ as our best forecast of this future probability. The estimated expected cost of d_i, $\hat{C}_i$, is then given by

$$\hat{C}_i = \sum_{j=1}^{k} C_{ij}\hat{p}_j$$

Table 10.4 Calculating decision costs (cost table)

Forecast events	Events occurring E_1		E_j		E_k	Decision	Estimated expected costs $\hat{C}$
E_1	C_{11}	...	C_{ij}	...	C_{1k}	d_1	$\hat{C}_1$
$\vdots$						$\vdots$	$\vdots$
E_i	C_{i1}	...	C_{ij}	...	C_{ik}	d_i	$\hat{C}_i$
$\vdots$						$\vdots$	$\vdots$
E_k	C_{k1}	...	C_{kj}	...	C_{kk}	d_k	$\hat{C}_k$
Forecast probabilities	$\hat{p}_1$	...	$\hat{p}_j$	...	$\hat{p}_k$		

Using a layout such as that shown in Table 10.4, the value of $\hat{C}_i$ can be calculated for the k possible decisions. It is then often a reasonable procedure to choose the decision d_h corresponding to the smallest estimated cost $\hat{C}_{min}$. If we wish to compare two different sets of forecasts, $\hat{p}$ and $\tilde{p}$, one could compare the corresponding sets of estimated expected costs $\hat{C}$ and $\tilde{C}$ One would certainly be interested to see whether they both lead to the same decision. Suppose, for example, there are only two possible decisions, d_1 and d_2; then it is clear that, on the basis of the table, when $\hat{p}_1$ is less than some value, p_0 which depends on the costs, we will reach decision d_2 and conversely when $\hat{p}_1 > p_0$. Thus any forecast will lead to the same decision provided it lies on one particular side of p_0. Consider another example of the use of these expected cost tables. Suppose that the costs associated with decision d_i when E_j occurs are constant, C say, for all $i,j(i \neq j)$. If, also, the costs associated with the correct decision $(i = j)$ are all at some constant value less than C, then the best decision is the one that corresponds to the most probable event as indicated by $\hat{p}$. This if $\hat{p}_3$ is the largest probability, d_3 is the decision made.

References

Brown, R. G. (1963). *Smoothing, Forecasting and Prediction of Discrete Time Series*. Prentice-Hall. Englewood Cliffs, New Jersey.
Chorofas, D. N. (1965), *Systems and Simulation*. Academic Press, New York.
Harrison, P. J. and Stevens, C. F. (1971). A Bayesian approach to short-term forecasting. *Op. Res. Quarterly*, **22**, No. 4, 341–362.
Theil, H. (1965). *Applied Economic Forecasting*. North-Holland Publishing Co., Amsterdam.

Chapter 11

Multivariate models

11.1 Introduction

Most people involved in forecasting have to forecast more than one variable. They thus in practice have to carry out multivariate (many variable) forecasting. For example, a sales department will usually be involved in forecasting the sales of many different products and also of the various versions of certain products that are made in different sizes, colours, etc. Usually each variable is forecast independently. There are many situations where this is done because it is expedient rather than natural. It is far simpler to treat each series separately than to get involved in the complexities of real multivariate forecasting. Certainly multivariate forecasting is a fairly sophisticated subject and most of it is beyond the scope of this book. For an advanced study see Hannan (1970) and Robinson (1967). In this chapter we will aim at introducing some of the ideas of the subject without getting too involved in mathematics. However, there is a real need to recognize the possible value of multivariate forecasting. Consider, for example, the following situation.

In certain markets price is controlled by the market rather than the seller. In such situations price and amount sold both vary and require forecasting. Clearly, each affects the other and to try and forecast each, ignoring the other, would lead to poor forecasts. To obtain reasonable forecasts we must allow for the past behaviour *and* interrelationships of the prices and the sales.

When discussing univariate forecasting, we made frequent use of the mean square error as a forecasting criteria. When we have several variables, we will have a mean square error matrix. This is an array of terms of the mean square error type. Thus, if $e_1, \ldots, e_n$ are the forecast errors in the n variables being forecast, the mean square error matrix is

$$\begin{pmatrix} E(e_1^2) & E(e_1 e_2) & \ldots & E(e_1 e_n) \\ E(e_2 e_1) & E(e_2^2) & \ldots & E(e_2 e_n) \\ \ldots & \ldots & \ldots & \ldots \\ E(e_n e_1) & E(e_n e_2) & \ldots & E(e_n^2) \end{pmatrix}$$

As this consists of many elements, we cannot regard it as a single criteria. We can imagine forecasting methods that will make some elements satisfactorily small at the cost of having others large. It is clearly very useful, however, to have a single criteria, and this is provided by the sum of the diagonal elements of the matrix:

$$I = E(e_1^2) + E(e_2^2) + \ldots + E(e_n^2)$$

This is called the trace of the matrix. If we imagine an n dimensional space in which the point $X = (x_1, \ldots, x_n)$ is the observed value of our n variables and $F = (\hat{x}_1, \ldots, \hat{x}_n)$ is the forecast value, then (by the theorem of Pythagoras in n dimensions) I is the mean square distance between F and X. Seeking to make this small is thus a reasonable approach to constructing a single criteria for multivariate forecasting.

11.2 Generalizing exponential smoothing*

So that we do not get too involved with mathematics, let us consider a bivariate forecasting situation with pairs of observations $(x_1, y_1), \ldots, (x_t, y_t)$. In ordinary exponential smoothing we discussed the use of the error correction form (section 4.2) for which

$$\tilde{x}_t = \tilde{x}_{t-1} + be_t$$

where e_t is the latest error, $x_t - \tilde{x}_{t-1}$, and $b = 1 - a$ is a forecasting parameter. The natural generalization of this is

$$\tilde{x}_t = \tilde{x}_{t-1} + b_{11} e_{x,t} + b_{12} e_{y,t}$$
$$\tilde{y}_t = \tilde{y}_{t-1} + b_{21} e_{x,t} + b_{22} e_{y,t}$$

where $\tilde{x}_t$ and $\tilde{y}_t$ depend on the errors, $e_{x,t}$ and $e_{y,t}$, of the last forecasts of both x_t and y_t. The constants b_{11}, b_{12}, b_{21} and b_{22} are the forecasting parameters. If this is written using matrix notation, its analogy with the single variable situation is clear. Thus

$$\begin{pmatrix} \tilde{x}_t \\ \tilde{y}_t \end{pmatrix} = \begin{pmatrix} \tilde{x}_{t-1} \\ \tilde{y}_{t-1} \end{pmatrix} + \begin{pmatrix} b_{11} b_{12} \\ b_{21} b_{22} \end{pmatrix} \begin{pmatrix} e_{x,t} \\ e_{y,t} \end{pmatrix}$$

This method is the method of multivariate exponential smoothing. It can be supported on similar grounds to ordinary exponential smoothing. Thus

(a) It corresponds to an exponentially weighted average of the observations with weights tending to zero in the far past (provided the b values satisfy certain conditions).

(b) Multivariate exponential smoothing is optimum, with appropriately chosen b values, for a set of models analogous to those for which ordinary exponential smoothing is optimum. For example, exponential smoothing is optimum for the model defined by

$$x_t = \mu_t + \epsilon_t$$

$$\mu_t = \mu_{t-1} + \eta_t$$

If x_t, μ_t, ϵ_t and η_t are now regarded as vectors, this describes a multivariate process for which multivariate exponential smoothing gives the optimum mean square error predictor, see Jones (1966) and Kalman and Bucy (1961). The parameters b in this situation depend on the variance—covariance properties of the ϵ and η sequences. Jones gives a fairly direct way in choosing appropriate values of these parameters.

11.3 Multivariate stochastic models

In Chapter 7 we considered how to forecast in situations modelled by a number of different types of stochastic process. All these processes can be generalized to cover analogous multivariate situations. The aim here is simply to indicate how these processes can be generalized and how the forecasts can be obtained for such multivariate situations. Consider, first, the generalization of a simple autoregressive process

$$x_t = \phi_1 x_{t-1} + \epsilon_t$$

Let η_t be the random variable for the y process and suppose that the present x_t and y_t depend on the past values of both the x and y sequences. Then we have

$$x_t = \phi_{11} x_{t-1} + \phi_{12} y_{t-1} + \epsilon_t$$

$$y_t = \phi_{21} x_{t-1} + \phi_{22} y_{t-1} + \eta_t$$

as a bivariate first-order autoregressive process. The generalization to higher orders simply involves introducing terms in x_{t-2}, y_{t-2}, etc.

We now turn to the moving average process

$$x_t = \epsilon_t - \theta \epsilon_{t-1}$$

and note that the latest ϵ and η in the autoregressive model only influence their respective x and y values. The natural generalization for the moving average process is thus

$$x_t = \epsilon_t - \theta_{11}\epsilon_{t-1} - \theta_{12}\eta_{t-1}$$

$$y_t = \eta_t - \theta_{21}\epsilon_{t-1} - \theta_{22}\eta_{t-1}$$

In generalizing both the above models we can increase the number of x or ϵ, or y or η, terms in the expression for both x_t and y_t. There are thus four numbers required to define the order of the process. If we introduce vector and matrix notation putting

$$\mathbf{x}_t = \begin{pmatrix} x_t \\ y_t \end{pmatrix}, \quad \boldsymbol{\epsilon}_t = \begin{pmatrix} \epsilon_t \\ \eta_t \end{pmatrix}$$

$$\boldsymbol{\theta} = \begin{pmatrix} \theta_{11} & \theta_{12} \\ \theta_{21} & \theta_{22} \end{pmatrix}, \quad \boldsymbol{\phi} = \begin{pmatrix} \phi_{11} & \phi_{12} \\ \phi_{21} & \phi_{22} \end{pmatrix}$$

Then the above models become

$$\mathbf{x}_t = \boldsymbol{\phi}\mathbf{x}_{t-1} + \boldsymbol{\epsilon}_t$$

and

$$\mathbf{x}_t = \boldsymbol{\epsilon}_t - \boldsymbol{\theta}\boldsymbol{\epsilon}_{t-1}$$

and the multivariate forms are directly analogous to the univariate forms. A consequence of this is that much of what was said about the univariate models in Chapter 7 carries over to the multivariate models. In particular, the optimum forecast, in the sense of section 11.1, is the conditional expectation of the future multivariate observation, given all past and present observations. Thus for the autoregressive model the predictor is $\hat{\mathbf{x}}_{t+1} = \boldsymbol{\phi}\mathbf{x}_t$, and for the moving average model it has the form $-\boldsymbol{\theta}\boldsymbol{\epsilon}_t$. As before, in both cases the one-step-ahead prediction error $\mathbf{e}_t = \mathbf{x}_{t+1} - \hat{\mathbf{x}}_{t+1}$ is equal to $\boldsymbol{\epsilon}_t$. Clearly, one can also extend these ideas to mixed autoregressive moving average models and to models in which some forms of trend, etc., can be differenced out. As a quick example, we note that if the model takes the form:

$$\mathbf{x}_t = \mathbf{x}_{t-1} + \boldsymbol{\epsilon}_t - \boldsymbol{\theta}\boldsymbol{\epsilon}_{t-1}$$

all the calculations given for the univariate model at the end of section 7.5 follow through. This gives as optimum predictor

$$\hat{\mathbf{x}}_{t+1} = \hat{\mathbf{x}}_t + (\mathbf{I} - \boldsymbol{\theta})\mathbf{e}_t$$

where $\mathbf{I}$ is a matrix of ones, which is exactly the form of multivariate exponential smoothing given in the previous section.

References

Hannan, E. J. (1970). *Multiple Time Series.* John Wiley and Sons, New York.

Jones, R. H. (1966). Exponential smoothing for multivariate time series. *J. Roy. Statist. Soc. B,* 28, 241—251.

Kalman, R. E., and Bucy, R. S. (1961). New results in linear filtering and prediction theory. *J. of Basic Eng.,* 83, D 95—108.

Robinson, E. (1967). *Multichannel Time Series Analysis with Digital Computer Programs.* Holden-Day Inc., San Francisco.

Chapter 12

Forecasting methods and models

12.1 Forecasting methods

In previous chapters we have examined a number of different models and discussed various forecasting methods that might be appropriate when these models describe, either globally or locally, a particular situation. Our classification has been based on models rather than methods. Table 12.1 gives a classification of some of the main methods that we have used to estimate the model parameters as the basis for extrapolating into the future. These fall into two main classes:

(a) those based on classical statistical methods of estimation, such as least squares or on simple modifications of these, such as those involving the use of the exponential discounting factor.
(b) those developed for particular situations, such as non-linear situations, in a much more intuitive and *ad hoc* fashion. Of these methods those which produce recurrence relations for updating the estimates of the parameters form an important special class.

Table 12.1 gives a summary of the advantages and disadvantages of these various methods.

12.2 Model building and identification

The approach of model building has been the underlying feature of the methods discussed in previous chapters. Though we devoted Chapter 2 to a discussion of models, there are a number of implications from the topics studied in later chapters that need comment. In any real situation it is unlikely that we can only talk about *the* model for that situation. We are usually in a position to reasonably use a number of different models. *The truth* in a situation is invariabliy too complex for anyone to grasp, assuming even that it exists. Different models will pick up different features of 'the truth'. We may model our data by a trend

180

Table 12.1 Forecasting methods

Method	Type of model for which method is most appropriate	Examples in previous chapters	Advantages	Disadvantages
Least squares	Global model simplest for linear models	5.1.2 7.1 9.2.2	Commonly available in computer packages; properties of forecasts simply derived, e.g. if model valid, obtain prediction intervals from least squares theory	Strongly dependent on well-behaved data and continuance of the assumed stable structure
Discounted least squares	Local model simplest for linear models	5.2.2 6.2 7.1 8.3 9.2.2	Is robust method that follows slow changes in model parameters; with exponential discounting, a^r, contains least squares as a special case — thus, in theory, cannot give worse forecasts if chose best a	As least squares is strongly effected by deviant observations, requires choice of the forecasting parameters (a)
Intuitive	Methods derived by intuition to give simple estimators for non-linear models, local or global models	5.1.1 5.2.3 8.2 9.2.3 9.2.4	Designed to be simple to understand and to calculate	The properties of the estimators obtained are often not known and may be poor
Recurrence	Local models, linear or non-linear	4.2 5.2.1 8.2	Designed to be simple to understand, calculate and update as new data are obtained; often are approximations to least squares and discounted least squares methods; are usually fairly robust methods	Usually require choice of forecasting parameters; often one forecasting parameter for each model parameter; properties of the estimators obtained are often not known

model, by a regression model on some economic variable or by a model involving the structure of the market. These models may overlap in some aspects; they may contradict each other in other aspects. We will certainly compare these models. We may find that one gives better forecasts over the last few years, but we would be unwise to discard the rest unless they clearly do not fit the situation at all. Each model gives its own insights into the total situation. Often the long-term value of constructing a model lies not so much in its forecasting ability, which may be poor and lead to its rejection, but in the insight into the data and the structure of the situation developed during the model-building exercise.

In Chapter 1 we classified forecasting methods roughly into three groups: (a) intuitive methods, (b) causal methods and (c) extrapolative methods. Though we have concentrated on methods in the third class, it should be noted that both of the other methods make use of model building. Clearly, the causal methods are essentially based on constructing a model that shows the causal structure of the situation. Most regression models and the more elaborate econometric models come under this category.

The problem of identification is that of identifying the model to use for a set of data. This is an aspect of statistics that lacks a formal theory. One proceeds in terms of graphical methods and of calculating appropriate statistics. In terms of practical forecasting the situation gets even more imprecise since for reasons of cost, availability of computer packages, ease of application, etc., one may finally opt for a model that is not the best possible one for describing past data. The process is rarely one of finding the best model; the chances are that we could always go on improving our models provided that we had nothing else to do. We usually start with a limited class of models and choose the best fitting of these. The models discussed in previous chapters give a choice of those most commonly found to be useful in forecasting. Table 12.2 lists a number of these models with an indication of what to look for in seeking to identify them.

The identification procedures described in Table 12.2 are mostly fairly straightfoward. The one exception is the identification of stochastic models. Several computer programs are now available that produce the relevant information for identification, but the interpretation of this information, to identify the appropriate model, still requires considerable skill.

The approach to identification in Table 12.2 is that of looking at appropriate plots and statistics. Another approach, though one which naturally combines with this, is to start with an assumed model and to examine its appropriateness. There are two ways of doing this. One is to start with a very sophisticated model that contains within it a number of simpler models. For example, a model with a polynomial trend and

Table 12.2 Identification methods

Model	Method

(a) Constant mean
$x_t = \mu + \epsilon_t$
	Plot x_t against t.
	Estimate μ by $\bar{x}$ and plot on graph.
	Calculate residuals $\hat{\epsilon}_t = x_t - \bar{x}$.
	Check these for: random scatter above and below zero,
	                 lack of independence (calculate
	                    autocorrelations),
	                 lack of constant spread.
	(See also methods of sections 12.3 and 15.2.)

(b) Linear trend $x_t = \alpha + \beta t + \epsilon_t$	Plot x_t against t. Fit a line by least squares and add to graph. Calculate residuals $\hat{\epsilon}_t = x_t - \hat{\alpha} - \hat{\beta}t$. Check these as above. Moving average of residuals + plot of residuals will indicate deviation from linearity.
(c) Exponential growth $x_t = \alpha\, e^{\beta t} + \epsilon_t$	Plot $\log x_t$ against t or x_t on logarithmic graph paper; this should give result as above. Fit by methods of Chapter 9. Residuals may not have constant spread.
(d) Seasonal models	Plots as in Chapter 8; see also plots of residuals and forecast errors as in section 15.2.2(e) and (i).
(e) Stochastic models	Calculate and plot using at least 50 observations.

Partial-autocorrelation[a]	+ Autocorrelation	→ Model identified
Zero after pth lag	+ Decaying values	→ Autoregressive process of order p
Decaying values	+ zero after qth lag	→ Moving average process of order q
Decaying values	+ Decaying values	→ Mixed AR—MA process

[a] For definitions see Appendix A.

several seasonal terms or a high-order autoregressive—moving average model might be appropriate. These models would be fitted to the data, then any terms in the model that made little contribution to the quality of it, i.e. whose coefficients were small, would then be neglected. This procedure is often called overfitting and clearly requires the availability of a large quantity of data. The second approach is to start with an appropriate simple model. The forecast errors, or residuals, are then carefully studied to see whether they show any remaining structure, e.g. autocorrelation or periodic behaviour. If such a structure is found and a suitable model fitted to it, then a combined model can be created. Methods for the analysis of errors and, equivalently, residuals are discussed in Chapter 15. If no structure is found in the errors, then our chosen model can be taken as satisfactory. Thus we have two approaches, one based on breaking down a complex model, the other on building up a simple one.

12.3 Model testing*

Forecasting is essentially a dynamic activity and one is usually concerned with questions as to whether the method currently used is still successful. Problems of this nature are about the quality of one's forecasting and are termed quality control problems. These are discussed at some length in Chapter 16. Often forecasting methods are applied almost as soon as any data are available and the problem of identifying and testing some suitable model is tackled on the basis of making, one hopes, inspired guesses and then applying quality control methods to test whether one is right. A consequence of this is that methods for formally testing models on an adequate set of data are not used in statistical forecasting as frequently as they are in other branches of statistics. None the less, if one has a set of data whose structure is not obvious, then there is a wide range of tests that one can use to test the fit of the more common models. These tests are also appropriate to a study of forecast errors. If forecasting is successful, the forecast errors should be a purely random sequence. If there are any correlation or seasonal features left in the error sequence, then it should be possible to improve on the forecasts. The aim of this section is to indicate briefly the nature of some useful tests. It will be assumed that the reader is familiar with the main features and forms of significance tests. Almost all introductory texts in statistical theory cover significance tests.

There are three main areas in which such tests might be useful to the forecaster.

(a) Testing for randomness

Here the question is whether there is any structure in the data at all. Kendall (1973) gives a number of different tests of randomness. Table 12.3 describes and illustrates one such test.

(b) Testing for trend

Table 12.4 illustrates two approaches to testing for a trend in the data. In part (a) a trend model is fitted and the test is whether the slope is significantly different from zero. In part (b) the data is used directly to derive a suitable test statistic. Notice that here the test statistic is designed to be particularly sensitive to trends. Thus in the example a trend is detected and the hypothesis of randomness rejected, whereas

Table 12.3 A turning point test of randomness

1. Definition

   A turning point is an observation that is greater than both its neighbours or less than both its neighbours.

2. Examples

   + is a turning point.

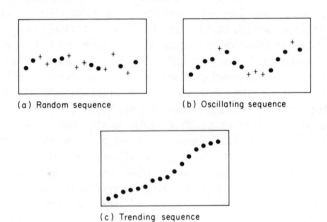

(a) Random sequence   (b) Oscillating sequence

(c) Trending sequence

3. Formulae

   n = number of observations

   p = number of turning points

   μ_p = expected number of turning points = $\frac{2}{3}(n-2)$ for random sequence

   σ_p = standard deviation of p = $\sqrt{\{(16n-29)/90\}}$

   $u = (p - \mu_p)/\sigma_p$ = standardized value of p

4. Test

   If data are from a random sequence u has, approximately, a standard normal distribution. Thus reject hypothesis of randomness if $|u| > 1.96$ for 5% significance test.

5. Example

Data	9.1	8.3	7.2	10.3	9.5	10.4	10.5	10.1	9.7	10.6
Turning points			+	+	+		+		+	

   $n = 10$, $p = 5$, $\mu_p = 5.3$, $\sigma_p = 1.2$, $u = 0.28$

   The observed u is near the centre of the standard normal distribution so the data is consistent with the hypothesis of randomness.

Table 12.4 Two tests of trend

(a)

1. Formulae (see Chapter 5 for detailed notation)

n = number of observations

$\hat{\beta}$ = least squares estimator of slope (β)

$\hat{\sigma}^2$ = estimated variance = $\Sigma\hat{\epsilon}^2/(n-2)$

$\hat{\sigma}_\beta^2$ = estimated variance of $\hat{\beta}$ = $\hat{\sigma}^2/\Sigma t'^2$

$t = (\hat{\beta} - \beta)/\hat{\sigma}_\beta$ = standardized value of $\hat{\beta}$

2. Test

If there is no underlying trend in the situation, $\beta = 0$. The statistic $t = \hat{\beta}/\hat{\sigma}_\beta$ can then be tested using a t-test.

3. Example (from Table 5.2)

$n = 11$, $\hat{\beta} = 1.882$, $\hat{\sigma}^2 = 2.546$, $\Sigma t'^2 = 110$, $\hat{\sigma}_\beta = 0.023$, $t = 81.8$, which is clearly highly significant, i.e. there is a clear trend in the data.

(b)

1. Formulae

n = number of observations

P = number of pairs of observations, not necessarily adjacent, in which the later one is greater than the earlier one

μ_P = expected value of $P = \frac{1}{2}n(n-1)$ for random sequence

$r = \dfrac{4P}{n(n-1)} - 1$; $\quad r = \begin{cases} +1, \text{ steady upwards trend} \\ -1, \text{ steady downwards trend} \end{cases}$

σ_r = standard deviation of $r = \sqrt{\{2(2n+5)/9n(n-1)\}}$

2. Test

If data are from a non-trending random sequence, r/σ_r has, approximately, a standard normal distribution. Thus reject hypothesis of non-trending randomness if $|r/\sigma_r| > 1.96$ for 5% significance test.

3. Example

Data	9.1	8.3	7.2	10.3	9.5	10.4	10.5	10.1	9.7	10.6
Number of points to left with lower value	0	0	0	3	3	5	6	4	4	9

$P = 34$, $r = 0.51$, $\sigma_r = 0.25$, $r/\sigma_r = 2.05$

The observed value r/σ_r is just significant at the 5% significance level; so there is some evidence of a trend, but more data would be useful before reaching a firm conclusion.

Table 12.5 An analysis of variance test of seasonality

(a) Form of data and calculations

	Season j			
	1	2	r	
Period 1	x_{11}	x_{12}	x_{1r}	
2	x_{21}	x_{22}	x_{2r}	
t	x_{t1}	x_{t2}	x_{tr}	Totals
Totals	S_1	S_2	S_r	ΣS_j
Sum of squares (e.g. $x_{11}^2 + x_{21}^2 + \ldots + x_{t1}^2$)	SS_1	SS_2	SS_r	ΣSS_j
Number of periods data (usually $= t$)	n_1	n_2	n_r	N
	S_1^2/n_1	S_2^2/n_2	S_r^2/n_r	$\Sigma S_j^2/n_j$
Sum of squared deviations (i.e. $SSD = SS - S^2/n$)	SSD_1	SSD_2	SSD_r	ΣSSD_j
Degrees of freedom	$n_1 - 1$	$n_2 - 1$	$n_r - 1$	$n - r$
Notation	$S_j = \sum_i x_{ij}, \; \bar{x}_{.j} = S_j/n_j, \; SS_j = \sum_i x_{ij}^2,$			
	$\bar{x}_{..} = \sum_j \bar{x}_{.j}/r$			

Variance analysis

Total sum of squares
$$\sum_i \sum_j (x_{ij} - \bar{x}_{..})^2 = \Sigma SS_j - \frac{(\Sigma S_j)^2}{N} \quad (1)$$

Between season sum of squares
$$\sum_i n_j (\bar{x}_{.j} - \bar{x}_{..})^2 = \Sigma S_j^2/n_j - \frac{(\Sigma S_j)^2}{N} \quad (2)$$

Within season sum of squares
$$\sum_i \sum_j (x_{ij} - \bar{x}_{.j})^2 = \sum_j \Sigma SSD_j = (1) - (2) = (3)$$

(b) Analysis of variance table and test

Source of variation	Sum of squares	Degrees of freedom	$(S \text{ of } S)/(d \text{ of } f)$	Test statistic
Between seasons	(2)	$r - 1$	$S_b^2 = (2)/(r - 1)$	
Within seasons	(3)	$N - r$	$S_w^2 = (3)/(N - r)$	$F = S_b^2/S_w^2$
Total	(1)	$N - 1$		

If there is no difference between the seasons, then S_b^2/S_w^2 has an
F-distribution with $(r - 1, N - r)$ degrees of freedom. Thus if F
is significantly larger than one, we reject the hypothesis that
there is no seasonal effect.

(c) Example: forecast errors (rounded) in quarterly forecasts

		Quarter				
		1	2	3	4	
Year	1	−10	2	9	−6	
	2	−3	−1	4	0	
	3	−4	5	1	−3	
	4	2	2	−6	−8	
	5	−2	4	7	4	Totals
Totals		−17	12	15	−13	−3
Sums of squares		133	50	183	125	491
Number of years		5	5	5	5	20
S^2/n		57.8	28.8	45.0	33.8	165.4
SSD		75.2	21.2	138	91.2	325.6
Degrees of freedom		4	4	4	4	16

Total sum of squares $\quad 491 - \dfrac{(-3)^2}{20} = 490.55$

Between season sum of squares $\quad 165.4 - \dfrac{(-3)^2}{20} = 164.95$

Within season sum of squares $\quad$ Difference $\quad = 325.6 \qquad$ checks with SSD above

Source of variation	Sum of squares	Degrees of freedom	$(S \text{ of } S)/(d \text{ of } f)$	F
Between seasons	164.95	3	54.98	
Within seasons	325.60	16	20.35	2.70
Total	490.55	19		

From the tables the value of F needs to be 3.24 to be significant at the 5% significance level. Thus there is no clear evidence that there is any seasonal variation left in the forecast errors.

for the same data the test of Table 12.3 was unable to detect non-randomness.

(c) Testing for seasonality

Table 12.5 illustrates how the method of analysis of variance, which is dealt with in many statistics texts, can be applied to testing for seasonality in data.

12.4 The use of judgement

There is a tendency for those, such as the writer, who approach forecasting from the statisticians' viewpoint to disapprove somewhat of 'judgement' as a means of obtaining a forecast. However, as we admitted in Chapter 1, even the most statistical of statistical forecasts involves the assumption of an underlying stable structure, at least locally. Whether this is so is a matter of judgement. It is a judgement based on a study of the factors in the situation. One might, for example, list the factors in the situation, government policy or consumer attitudes say, and judge whether they are likely to change sufficiently in the lead time to upset one's model. If the answer is yes, then one may have to produce a forecast by applying one's feel for the situation to adjust the mathematically derived forecast. The point, however, is that even if one leaves the statistical forecast alone *that* decision is based on judgement.

Forecasts are sometimes produced by judgement alone. Often this is done because there is insufficient time allowed for obtaining and using data in a statistical forecast. Sometimes the data may not exist at all or exist only in inadequate amounts. It may be that a situation is changing so rapidly that a statistically based forecast would be no use even as a guide. This does not mean, however, that studying the situation from a model-building point of view would not help in the forming of judgement. The very essence of producing good forecasts by the use of judgement is the careful study of the situation from as many viewpoints as possible.

On the basis of a feeling that more heads are better than one, many judgemental approaches are based on getting a consensus forecast from a number of individuals. These vary in name and type from the committee of area sales managers and the 'Jury of Executive Opinion' (Bassie, 1958) to the surveys of forecasts by relevant experts carried out by learned societies, banks, newspapers and magazines. There is a temptation here to argue from the fact that an average, $\bar{x}$, is a better estimate of the underlying mean than a single observation. In fact one can argue that as some forecasters will be better than others, one should use a weighted mean of the individuals forecasts, with weights increasing with the individuals past successes at forecasting. If all the experts produce unbiased forecasts, this procedure would certainly be reasonable. However, it is likely that an unexpected downswing in, say, sales would be unexpected to all the group and the consequent bias would appear in the consensus forecast as in the individual. However, this does not rule out the method but rather points out that it has no magic ingredient. The problem of how to combine the individual forecasts has been met with a variety of solutions. The simplest is the type of averaging operation suggested above. This could be modified to allow for the bias that will almost certainly occur in the forecasts of some members of the group. This occurs because of the way in which the forecast will influence their part of the organization; a high forecast will make life harder for production men but possibly more lucrative to salesmen. Over a period of time the magnitudes of these biases could be assessed and these forecasts adjusted accordingly.

Other methods seek to get the group to work as a team and arrive at a genuine consensus. This can be done by circulating initial forecasts, each with some indication of the reasoning behind its choice, on the basis of which each member revises his forecast. This is usually done anonymously. The feedback produced by the study and revision of forecasts reduces the spread of the forecasts. This is particularly so if those well above or below the average have to produce detailed arguments for keeping to their forecasts, that is if they wish to. Part of the value of this type of exercise is the investigation of relevant factors as seen from a number of different viewpoints. It is possible in this technique to introduce something corresponding to the weights in the previous method. In the initial circular, previous forecasts and actuals could be given for each of the forecasters. Thus people would begin to form a clearer assessment of their own and other people's biases and tendencies.

A major problem occurs with judgement forecasts that is not present to the same extent with forecasts based on models. With a statistical forecast, assuming stability, one can evaluate standard errors for the future observations, given the forecasts. If the same forecaster makes repeated forecasts on the same basis of information, then there will exist a set of errors with probably a clear statistical pattern. Where we are dealing with consensus forecasts using possibly different people and different lines of argument each time, it is likely that the errors over a sequence of forecasts will not show so much stability. If this is the case, then it is difficult to assess limits for the forecast. One could try and form a consensus opinion of the range in which the future observation will lie with given probabilities. Alternatively, the revised, or original, sets of individual forecasts could be treated as a set of random observations on a distribution with the consensus forecast as mean. This distribution could then be used to provide an error distribution. Experience would have to be built up to see how close the distribution of errors in forecasts corresponded to the distribution of forecasts made by individuals.

References

Bassie, V. L. (1958). *Economic Forecasting*. McGraw-Hill, New York.
Kendall, M. G. (1973). *Time Series*. Griffin Co. Ltd. London.

Part III

The forecasting process

The following chapters explore a range of problems and methods associated with the development and implementation of statistical forecasting methods.

Part III

Chapter 13

Data

13.1 Introduction

The quality of a forecaster's results cannot be better than the quality of his data. It is therefore worthwhile to devote a short chapter to looking more carefully at the forecaster's data. In section 13.2 we will look at the sources of his data and in section 13.3 we will examine the quality of his data. The 'quality' of data is almost impossible to measure or define, so we will avoid the issue by simply looking at situations where the 'quality' of the data is clearly low and seeing what might be done about it.

We will also look at some of the many ways in which data are adjusted for use in forecasting.

13.2 Sources of data

The sources of data depend on the type of variable being forecast. The sources for forecasting sales of furniture will be very different from those required for forecasting the number of school places needed for ten-year-olds in a town. There is no hope, nor need, to give a study covering all applications. Instead, we will concentrate on forecasting in the business area of application, as this is the most common and as it also illustrates a number of more general aspects.

In Chapter 2 we made a distinction between internal and external data, i.e. data obtained within the forecaster's own firm and data obtained from outside. Let us look at internal data first. It is essential to the forecaster's work that, as part of the firm's procedures, appropriate data are collected and recorded. He should look very carefully at how data are initially obtained, combined and circulated so that he can obtain data of the highest quality possible as early as possible. Where the forecaster is starting from scratch and trying to get together past data, a lot depends on knowing the firm and its staff very well. In one firm the sales department discovered that, unfortunately, they could not trace any records of an important piece of information about orders for any time before the previous year. This was casually mentioned to the production manager who thereupon produced a sheet of paper on which he had noted this information every month for the last ten years 'for general interest'.

When handling the firm's own data it tends to be assumed that the firm knows all about it. Where the data originate from the 'customer'. there are often many aspects that are not apparent from the numbers themselves. What is the customer buying for? What is his buying policy? These are very relevant questions. A customer who buys material may change his use of the material and thus possibly his future buying pattern. A customer who buys both for direct use and for stock may well produce misleading orders in times of change. The more the firm knows about its customers the better. Where there are a few major customers, relevant information may be provided willingly by the customer if he can see that it could benefit him in terms of, say, more reliable deliveries. Alternatively, he could be persuaded to provide information by way of appropriate offers. Where there are a lot of customers the problem is more complex. The natural procedure is to use samples. This, however, is very expensive, but some consulting firms use such samples to study large numbers of products and can thus spread the costs. The information on buying patterns, etc., provided by such firms can often be very valuable, particularly in relation to new products and to assessing who buys the products.

Forecasters in the past have often been defeated by the lack of data. The current developments towards the creation of effective data bases for firms should greatly ease the forecaster's problems in this context. As well as internal information, an effective information system for the forecaster should contain information about the external environment. Pearce (1971) gives a study of such systems within the marketing context.

As well as the problems of data shortage, the forecaster is sometimes faced with problems of data abundance. In a stock control situation of a manufacturing firm there may well be many thousands of items requiring a weekly forecast. In such a situation a detailed study of models and methods for forecasting each item is out of the question. Probably the best procedure is to seek to classify the items into groups on the basis of criteria such as type of use, value per unit and magnitude of use. A random sample of items from each group can then be used as the basis for examining forecasting methods. It would be sensible in this situation to use some items in each group as the basis for choosing methods and parameters and other items for checking the effectiveness of these methods. Clearly, the robustness of the methods is of great importance when they are used to forecast series that have not themselves been investigated.

External data for business forecasting come under roughly two headings:

(a) data for similar products, or generally for the same industry;
(b) data concerning the general business environment and economy of the country.

Considering industry data sources first, these are mainly the institutions or societies representative of the industry. Some industrial magazines publish a wide set of statistical data on sales, etc., for the industry as a whole. These data frequently come from government sources and are based on returns made by individual firms to central government. How much detailed statistics has to be supplied to central government by firms varies from country to country. Some industries have schemes by which firms can join an organization to which they supply confidential information on many things from sales to wages. This information is then compiled in such a way that the data from individual firms cannot be identified and the statistical report is sent on a monthly basis to all firms in the organization. This approach enables firms to have useful information very quickly. Mention should also be made of commercial consulting agencies and research organizations who are either associated with particular industries or make occasional studies of these industries. These organizations often produce statistics on an industry of relevance to forecasting.

National data provide information to assist in the general judgemental background of forecasting as well as in the statistical techniques. These data include statistics on the economy as a whole, on particular parts of the economy, including data on geographical regions of the country. As well as this general information, some agencies select and publish series that are particularly useful as lead series, or as indicators of turning points and cycles. As all these publications vary from country to country, there is little point here in listing large numbers of publications. In the books by Silk and Curley (1970) and Woy (1965) the reader should find information to indicate how to look further.

13.3 The quality of data

Let us turn now to the quality of the data used in forecasting and list some of the things that can be wrong with such data.

(a) The data may be unavailable or late. For example, in regression methods substitute variables have to be used where an intuitively important regressor variable is either not known or is not published until after the deadline for producing the forecast. The substitute variable may be an estimate of the missing one based on sample information or it may be some other variable that has a high correlation with it. There are occasions when information can be speeded up. Certainly government statistical agencies have been working hard in this direction in recent years. In a number of cases firms have greatly improved their very short-term forecasting for stock control use by replacing mailed information by telex or by remote computer terminals.

(b) The data may be inaccurate. Before arriving on the forecaster's desk, the data will have been collected and processed in various ways. What can be done about inaccuracy will depend on the situation. A number of examples will suffice to point to the value of looking at this aspect of the data. A firm based its forecasts and its long-term production schedules on annual estimates by its customers, who were large and few in number. Originally the data produced bore only a remote relation to the final orders. At a later stage these were significantly improved by offering the forecast amount at a discount. The firms therefore found it worthwhile to think more carefully about their annual estimate. A similar procedure has been adopted with sales forces where financial incentives have been introduced to encourage accurate estimation by the salesmen of their future sales. Where data are copied in transmission, errors are almost certain to creep in. Where forecasting is done by computer on a large scale, it is worthwhile including in the programmes simple tests to check that numbers have the right order of magnitude and the right form.

13.4 The adjustment of data

When one talks to practitioners of forecasting it is often apparent that, though they will say that they use some standard method, they have their own particular versions appropriate to their own particular situations. The most common variations are linked with the idea of making adjustments of various types to the data or the forecasting formulae. In the first of these types the raw data are altered in some fashion before being put in the forecasting formula. In the second, the raw data are used but adjustments are made to the formula to allow for some features of these data. The aim of this section is to briefly discuss and reference some of these adjustment procedures.

13.4.1 Adjusting for time period

Most forecasting methods are based on assuming that data arrive at fixed regular intervals. This assumption may not be valid and the adjustment is made to produce a new set of data for which the assumptions are more nearly valid. For example, in a regression problem, the regressor variables may be yearly figures. If one is published quarterly, then clearly the quarterly figures can be added together to obtain annual figures. The problem, of course, comes when the situation is the other way round. Quarterly figures may be obtainable for all the variables save one, which has only annual data. In this case it may be necessary to forge quarterly figures for the variable in question. The only necessary condition on these figures is that they add up to the correct yearly totals, x_t. One approach is to make each quarterly figure a weighted average of three yearly totals. Thus the quarterly figures for year t, y_t^I, y_t^{II}, y_t^{III}, y_t^{IV}, are given by

$$y_t^I = ax_{t-1} + ex_t + dx_{t+1}$$

$$y_t^{II} = bx_{t-1} + fx_t + cx_{t+1}$$

$$y_t^{III} = cx_{t-1} + fx_t + bx_{t+1}$$

$$y_t^{IV} = dx_{t-1} + ex_t + ax_{t+1}$$

It will be seen that the weights a, b, c, d, e, f are chosen to give a reasonable symmetry to the weighted averages. To choose actual values for a, b, c, d, e, f, we add to the above condition some further reasonable conditions. For example, if the quantities x_{t-1}, x_t, x_{t+1} follow a trend, y_t^I, y_t^{II}, y_t^{III} and y_t^{IV} ought to follow the same trend. Boot, Feibes and Lisman (1967), Ginsburgh (1973) and Lisman and Sandee (1964) give a fuller discussion with some appropriate formulae.

Another area of common occurrency is that produced by the use of unequal time periods. For example, a firm's monthly figures may be based on four five-week months and eight four-week months. To produce a set of comparable figures, we must adjust each monthly figure by multiplying by a weighting factor, say w_1, if there are four weeks in the month and w_2 if there are five weeks. To obtain comparable figures, if weekly sales are constant, say one unit per week, we must have

$$w_1 \cdot 4 = w_2 \cdot 5$$

We must also have a set of figures giving the correct annual total, so that

$$8 \cdot w_1 \cdot 4 + 4 \cdot w_2 \cdot 5 = 52$$

Solving these two simultaneous equations gives

$$w_1 = 13/12 \text{ and } w_2 = 13/15$$

Having multiplied each actual monthly figure by w_1 or w_2 and substituted the adjusted figure in the forecasting formula, we must remember to divide the forecast by w_1 or w_2, as appropriate, to get a proper forecast of the future value.

In the above situations we have had some basic time interval from which to make adjustments. It can occur that the data arise at completely irregular times t_1, t_2, ... t_k. The observations x_1, ..., x_k would usually represent the number of 'items' arising in the intervals $(0, t_1)$, (t_1, i_2), ..., (t_{k-1}, t_k), respectively. In this situation the under-lying variable of interest is the rate variable, the number of items arising per unit time. We would thus adjust each observation by dividing it by the length of its time intervals. The adjusted data, the rates, would thus be r_1, ..., r_k, where $r_k = x_k/(t_k - t_{k-1})$.

As forecasting methods are often based on discounted methods (undiscounted methods being included here as a special case with weights $w = 1$), we can use this type of data in the standard methods provided we use appropriate weights for the discounting. for example, if $t_k - h_j$ is a time in the middle of the interval (t_{k-j-1}, t_{k-j}), then the weight associated with r_{k-j} could be a^{h_i}. This would modify the use of exponential weights in a suitable fashion. An alternative procedure is to use standard forecasting formulae modified in an appropriate manner. By way of an example, consider the modification of exponential smoothing. The basic formula is

$$\tilde{r}_k = a\tilde{r}_{k-1} + (1-a)r_k$$

We now seek to modify this to allow for the irregular intervals. We would expect that r_k represents a better estimate of the rate for larger time intervals $t_k - t_{k-1} = T$, say, so that a choice of a that varied inversely with T would seem appropriate. One way of picking out such a relationship would be to vary a in such a way that the $\tilde{r}_k$ values have some convenient property. One suggestion is that if the rate follows a linear trend the lag in $\tilde{r}$ remains constant in spite of the variation in T. Suppose that the rate has a slope α and is μ at time t_k. The lag of $\tilde{r}$ behind r will be proportional to α and denoted by $c\alpha$. From the diagram of Figure 13.1 it is seen that the expected values of $\tilde{r}_k$ and $\tilde{r}_{k-1}$ are given by

$$E(\tilde{r}_k) = \mu - c\alpha$$

$$E(\tilde{r}_{k-1}) = \mu - \alpha T - c\alpha$$

The expected number of items arising in the time T will be the average rate times T, i.e. $(\mu - \alpha T/2)T$. Thus the expected value of r_k is

$$E(r_k) = \mu - \alpha T/2$$

Thus if we substitute these expected values in the exponential smoothing equation, we obtain

$$\mu - c\alpha = a(\mu - \alpha T - c\alpha) + (1-a)(\mu - \alpha T/2)$$

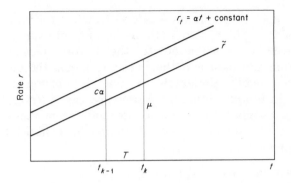

Figure 13.1 Trending rates

Solving gives $a = (2c - T)/(2c + T)$. Thus at each use of the recurrence relation a is recalculated from T. The choice of the best a is replaced by the choice of the best c.

   A final comment on the consideration of time scales is that even if one can obtain data at regular intervals it is not necessarily the best thing to do. As an example of this point, consider some attempts that have been made to model share prices (Mandelbrol and Taylor, 1967). The obvious time scaling is to take one unit as one day's trading, so that the x_t are the closing prices on the Stock Exchange. However, some days are slow days for a given share, with very little buying and selling. On other days as much stock is bought or sold as in the average week. On these latter days it is as though time speeded up for that particular stock. It is reasonable to try to scale t so that a uniform rate of buying and selling occurs. This is done by observing x_t at times called $t = 0, 1, 2, \ldots$, which correspond to the moments at which, say, 0, £10,000, £20,000, £30,000 pounds worth of stock has been traded, provided this information is available. This keeps a constant amount of trading, rather than a constant amount of physical time, between successive observations x_t.

13.4.2 Adjusting for known causes

   In many sets of data there are certain effects that are known to influence the raw data. Examples are

(a) holidays, as in sales from retail firms;
(b) the weather, as in sales of ice-cream;
(c) seasonal factors generally;
(d) strikes and other known hold-ups;
(e) stockouts.

Ways of dealing with these vary greatly. Some are pure intuition, where, for example, a guessed amount is added to allow for holiday lost sales. Others are fairly sophisticated methods, where, for example, parameters are introduced to correspond to the known factors, which are estimated from the data and then used to adjust the data. The latter methods are clearly preferable if one is dealing with repeated occurrences. In unique situations caused by, say, a strike, one is forced to make a guestimate on whatever grounds seem reasonable. As an example of the more mathematical approach, a model for ice-cream sales was expressed as:

$$x_{j,k} = \alpha + \beta t_j + \gamma r_j + \delta T_{j,k} + \epsilon_{j,k}$$

where $x_{j,k}$ is the sales in week j of year k, t_j is the temperature norm for that week, r_j is a seasonal term and $T_{j,k}$ is the average deviation of that week's temperature from the norm. A set of past data was used to estimate the parameters. A figure for adjusted sales for normal temperatures was then found by subtracting the estimated effects of $T_{j,k}$, namely $\delta T_{j,k}$, from $x_{j,k}$. Another common example is the use of seasonally adjusted data for regressor variables in regression forecasting. One general comment here is that though such adjusted data are often useful at the data analysis stage, they may not be so useful for forecasting. To see why this is so, note that the process works in three stages:

(a) A model for the data, involving the effects of interest, is fitted.
(b) The data is adjusted by eliminating these effects.
(c) The adjusted data is used with an appropriate forecasting model.

These three stages can usually be combined by incorporating the model for the data, *with* the effects, in the general forecasting model. Experience indicates that this more general model will usually provide better forecasts than the three-stage approach.

13.4.3 Transforming the data

A very common practice in forecasting is the application of simple transformations to the data before carrying out the forecasting. The two most common transformations are

$$y_t = \log_e x_t$$

$$y_t = x_t - x_{t-1}$$

The first of these transformations is commonly applied to reduce percentage changes to absolute changes, to reduce a growth model to a linear model or to change from a multiplicative to an additive seasonal model. The forecast of x_{t+1} is obtained as $e^{\hat{y}_{t+1}}$. As mentioned in the discussion of growth curves, an unbiased forecast of y_{t+1} will lead to a biased forecast of x_{t+1} when this transformation is used. It is possible to make allowances for such biases and this is discussed in section 16.2.

The second transformation is frequently used as a device for removing the effects of a wandering mean. If the mean μ_t is locally constant, then the differenced series y_t will not depend on μ_t, since it cancels between x_t and x_{t-1}. The forecast $\hat{x}_{t+1}$ will be $x_t + \hat{y}_{t+1}$.

There are obviously many transformations that might be useful in particular circumstances. Two questions need to be considered in deciding whether to use a transformation. The first is the obvious one as to whether better forecasts of x are obtained by the use of a transformation. The second question is whether there is a reasonable intuitive or theoretical justification for the use of the particular transformation. If we are to expect the transformation to lead to good forecasts in the future, we must have some reason to believe that it related to the underlying structure of the situation.

13.4.4 Adjusting for outliers

Most practical forecasting systems contain quality control procedures that will pick out values that are in some sense extreme. For example, the forecast error may be more than say 2½ standard deviations from the mean error. Such values are often referred to as outliers. The quality control procedure will examine such items to see if they are indicative of a real change in the underlying situation. Often, however, they are just isolated occurrences. In initial investigations of forecasting it is advisable to put some effort into seeking explanations for outliers. For example, in forecasting for production planning purposes a number of outliers were found to correspond to single exceptional orders which needed special production arrangements. The procedure was not therefore to try and forecast the outliers along with the rest, but to separate them as a special category. The result of this was to improve the quality of the forecasts made for the normal run of items.

Where outliers occur in an operational forecasting system, it is advisable to replace the outlier by its forecast or some other suitable less extreme value, otherwise future forecasts based on the outlier will be badly out. An indication of less subjective approaches to outliers, transients, was given in section 10.3.

References

Boot, J. C. G., Feibes, W., and Lisman, J. H. C. (1967). Further methods for the derivation of quarterly figures from annual data. *Applied Statistics*, 16, 65—75.
Ginsburgh, V. (1973). A further note on the derivation of quarterly data consistent with annual data. *Applied Statistics*, 22, 368—374.
Lisman, J. H. C., and Sandee, J. (1964). Derivation of quarterly figures from annual data. *Applied Statistics*, 13, 87—90.

Mandelbrol, B., and Taylor, H. M. (1967). On the distribution of stock price differences. *Opns. Res.*, 15, 1057—1062.

Pearce, C. (1971). *Prediction Techniques for Marketing Planners.* Assoc. Bus. Programmers, London.

Silk, L. S., and Curley, M. L. (1970). *A Primer on Business Forecasting.* Random House, New York.

Woy, J. B. (1965). *Business Trends and Forecasting—Information Sources.* Gale Research Co., Detroit.

Chapter 14

Adaptive methods and other extensions

14.1 Introduction

We have discussed thus far a fairly limited number of forecasting methods. The practicalities of forecasting often pose problems that are not adequately dealt with by these methods. The situation may change so rapidly that even our local model-building approach will not work. A 'fracture' may occur in which a model parameter suddenly changes its value. The situation may be highly complex and involve variables other than time. It may require the use of different approaches at different stages in the problem . It may be that such new problems will require new models and new methods. However, experience suggests that it is worthwhile to look at the methods that we already know and seek to modify and to extend them to meet the new situations. The aim of this chapter is to introduce some of the ways in which this is possible.

14.2 Adaptive methods

We consider in this section ways of getting our standard forecasting methods to adjust to rapid changes in parameter values. Clearly, if global methods are being used then, when a change has been found to occur, we must wait until enough data have been accumulated to enable a model to be refitted. Thus global methods are not very suited to this type of situation. Local methods are more suitable but they need to be adjusted to deal with rapid changes.

To see how this can be done we take, as our illustration, forecasts based on simple exponential smoothing as discussed in Chapter 4. On several occasions in past chapters we have observed that decreasing the smoothing constant, a, makes a forecast more sensitive to variation in the data, both random variation and that due to real changes in structure. Conversely, increasing the smoothing constant reduces the error variance, but also makes the forecast slower at following changes in structure. It is therefore natural to look into ways of changing the smoothing constant while forecasting, the general aim being to reduce it on occasions where there are changes in mean and to increase it when the situation is stable. There are several ways of doing this that have proved effective in practice. One possibility is that of using a subjective assessment of the stability of the situation as the basis for adjusting the smoothing constant. In many situations this is obviously a sensible approach. For example, if a large advertising campaign is to be undertaken, one would hopefully reduce the smoothing constant so that the forecasts would rapidly pick up any increased level of sales.

To develop some more objective ways of varying the smoothing constant, consider first how a single fixed value might be chosen. First we choose a set of alternative values for the smoothing constant a, e.g. 0.2, 0.4, 0.6, 0.8, 0.9, 1.0. Applying each of these in turn, we can calculate forecasts for our data and for each set of forecasts evaluate the appropriate forecasting criterion C, e.g. C = MSE. We finally choose the value of a that gives the smallest value of C. This procedure can be modified to enable the value of a to be adjusted as forecasting proceeds. We may adjust a at regular intervals by going through the above procedure each time, using only the data in the last interval (e.g. see Eilon and Elmaleh, 1970). Alternatively, we may re-evaluate a, using only recent data, whenever some quality control procedure indicates that a change may have taken place. Ideally we would like to be able to adjust a, if necessary, every time an observation is taken. One way of doing this is to evaluate some local forecasting criterion at only a limited number of values of a. We might, for example, fix on two values of a: a high value for stable situations and a low value for periods of change. The high value would normally be used, but the low value would be dropped to whenever the value of C indicated that it was necessary. Another approach which continually searches for the best a, but only requires the evaluation of C at three values of a, is illustrated in Table 14.1. Details of these and related methods are given by Chow

Table 14.1 A search for the best a

Values of C for given a	0.70	0.75	0.80	0.85	0.90	Best value of a
		75	63	79		0.80
		73	68	81		0.80
		71	70	80		0.80
		70	72	81		0.75
	75	69	73			0.75
Time	77	69	73			0.75

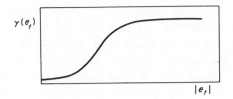

Figure 14.1   Forms for $\gamma(e)$

(1965), Roberts and Reed (1969), Roberts and Whybark (1973), Montgomery (1970) and Rao and Shapiro (1970).

Yet another approach is to make the smoothing constant a function of forecast errors. The method of Van Dobben de Bruyn (1964) does this by using simple exponential smoothing in the error correction form

$$\tilde{x}_t = \tilde{x}_{t-1} + \gamma(e_t)\, e_t$$

The smoothing constant is thus made a function of the latest error e_t. By choosing the form of the function $\gamma(e_t)$ to be like that in Figure 14.1, it is possible for the formula to behave like ordinary exponential smoothing with a high smoothing constant when e_t is small. When, however, large errors occur, the value of $\gamma(e_t)$ increases and these errors have a consequent large effect of the estimated mean. A danger here, of course, is overreaction and a consequent instability. The original work of de Bruyn allows for this by using a more elaborate form of the error correction approach. An alternative is to make the smoothing constant depend on the whole sequence of past errors, though obviously with emphasis on the latest values. We thus need to devise a function that either increases or decreases when the sizes of errors tend to get larger. One such function, that will be discussed further in Chapter 16, is defined by

$$|S_t| = |\,\text{smoothed error/smoothed absolute error}\,|$$

$$= |\,\tilde{e}_t/|\,\tilde{e}_t\,|\,|$$

The smoothed error gives a measure of bias in the errors. The smoothed absolute error gives a measure of variability with which to compare the bias. It can easily be shown that $0 \leqslant |S| \leqslant 1$ and it is clear the $|S|$ will equal one when all the errors have the same sign. The smoothing constant can thus be set equal to $1 - |S|$ or to some multiple of this, possibly with a restriction to limit $|S|$ below some maximum value. The restriction is sometimes introduced to stop the smoothing constant getting unrealistically small. Thus one might have

$$a_t = \begin{cases} 1 - c\,|S_t|, & |S_t| \leqslant S\max \\ 1 - c\,S\max, & |S_t| > S\max \end{cases}$$

The use of such an 'adaptive' parameter was introduced in Trigg (1964) and Trigg and Leach (1967) and the idea has been widely used e.g. Bamber (1969). One word of caution is necessary for the case where there are several parameters in the model. If the estimators of parameters such as slope are allowed to change rapidly owing to the use of an 'adaptive' smoothing constant, the forecasting method can become unstable and give wild forecasts. In these cases it is best to use a form of forecasting in which several smoothing constants can be used. That associated with estimating the underlying mean of the process can be made adaptive, but the other smoothing constants should be kept constant. By way of example, for forecasting a linear trend, Holt's method, as discussed in Chapter 5, would become

$$\hat{\mu}_t = \hat{\mu}_{t-1} + \hat{\beta}_{t-1} + (1 - a_t)e_t$$
$$\hat{\beta}_t = \hat{\beta}_{t-1} + ce_t$$

where a_t is obtained adaptively and c is a fixed smoothing constant.

The approaches above all adapt to changes in the model parameters. A more general alternative is to seek to adapt to changes in the model itself. In this approach the data obtained to date are used to select the most suitable model from a particular set of possible models. As another approach it is possible to consider models in classes and to always use the simplest model in the class that currently fits the situation. For example, the constant mean model is a special case of the linear trend for the slope $\beta = 0$. If in Holt's method, Chapter 5, we constantly update $\hat{\beta}$ but use the constant mean forecast unless $\hat{\beta}$ is significantly different from zero, then we can at least be adaptive between these two models.

14.3 Extensions to recurrence formulae

When exponential smoothing was first discussed, it was shown that when a^t can be neglected the exact form can be approximated by the following reccurrence relation:

$$\tilde{x}_t = (1 - a)x_t + a\tilde{x}_{t-1}$$

or, in words,

New forecast = $(1 - a)$ × new observation + a × last forecast

In section 5.2 we introduced an extension of the recurrence form to allow for a trend. This method was called Holt's method. It will be recalled that to use this we denote the current mean by μ_t and the slope by β_t. The model was thus

$$x_{t-h} = \mu_t - \beta_t h + \epsilon_{t-h}$$

the recurrence relations for mean and slope were

$$\hat{\mu}_t = (1 - a)x_t + a(\hat{\mu}_{t-1} + \hat{\beta}_{t-1})$$

and

$$\hat{\beta}_t = (1 - b)(\hat{\mu}_t - \hat{\mu}_{t-1}) + b\hat{\beta}_{t-1}$$

or, in words,

New estimate of mean = $(1 - a)$ × estimate based on new data

+ a × estimate of current mean based on last estimates of μ and β

New estimate of slope = $(1 - b)$ × estimate of slope based on change in estimated means

+ b × last estimate

The use of two recurrence relations introduces the use of two smoothing constants a and b.

The aim of this section is to illustrate how the use of recurrence relations can be extended to deal with more complex situations. For an example we consider forecasting sales in a situation where weather is a factor of importance. This particular feature occurs for many commodities, from ice-cream to electricity. The papers by Berrisford (1965) and Chen and Winters (1966) give specific examples of this type of analysis. Suppose that sales x_t for day t can be modelled by a constant underlying mean μ_t to which is added the effect of

(a) the deviation d_t of the temperature on that day from the norm for that day in the year, the norm being given by the average temperature over a number of years for that particular day in the year.
(b) the day in the week effect $W_i(i = 1, \ldots, 7)$, since, for example, sales on Saturday, $i = 6$, may be made higher than those on any other day.

The final model is thus

$$x_t = \mu_t + W_i + \alpha d_t + \epsilon_t$$

where α is a parameter, that we will assume known, measuring the increase in sales per degree deviation in temperature. In practice α would probably vary according to the time of year. As usual ϵ_t represents the random variation in the situation. Suppose at day t we have estimates of μ_t and W_i, denoted by $\tilde{\mu}_t$ and $\tilde{W}_t$, and also a weather forecast of d_{t+1}, denoted by $\hat{d}_{t+1}$. The forecast of x_{t+1} is then

$$\hat{x}_{t+1} = \tilde{\mu}_t + \tilde{W}_{i+1} + \alpha\hat{d}_{t+1}$$

When x_{t+1} and d_{t+1} have been observed, they may then be used to update $\tilde{\mu}_t$ and $\tilde{W}_{t+1}$ using recurrence relations. Consider $\tilde{\mu}_t$ first. From the model it is seen that if we have last week's estimate of $\tilde{W}_{i+1}$, denoted by $\tilde{W}_{i+1,\text{old}}$, a new estimate of μ is given by

$$x_{t+1} - \tilde{W}_{i+1,\text{old}} - \alpha d_{t+1}$$

Thus this new estimate, based on a single observation, can be put in a recurrence relation with the old estimate to give

$$\tilde{\mu}_{t+1} = (1-a)(x_{t+1} - \tilde{W}_{i+1,\text{old}} - \alpha d_{t+1}) + a\tilde{\mu}_t$$

a being the smoothing constant. This formula, as with all our previous recurrence forms, enables us to re-estimate μ_t in a very simple fashion, and also, being of exponential recurrence form, it will tend to follow local changes in μ_t. The same approach may now be used again to update any of the day of the week effects. Thus from the model, remembering that $\tilde{\mu}_{t+1}$ is now known, the quantity

$$x_{t+1} - \tilde{\mu}_{t+1} - \alpha d_{t+1}$$

gives an estimate of W_{i+1} based on the observed x_{t+1}. Thus again a recurrence relation for the W values can be constructed to give

$$\tilde{W}_{i+1,\text{new}} = (1-b)(x_{t+1} - \tilde{\mu}_{t+1} - \alpha d_{t+1}) + b\tilde{W}_{i+1,\text{old}}$$

where b is a second smoothing constant. If it is thought that α is a constant, without any seasonal change but with possible local variation, we could extend the above to produce a recurrence relation for α, namely

$$\tilde{\alpha}_{t+1} = (1-c)(x_{t+1} - \tilde{\mu}_{t+1} - \tilde{W}_{i+1,\text{new}})/d_{t+1} + c\tilde{\alpha}_t$$

where c is yet a third smoothing constant, which should be very close to unity.

In our initial introduction of the use of recurrence relations in forecasting we carried out three steps:

(a) Estimate the model parameters, say by discounted least squares.
(b) Express the estimators obtained in recurrence form.
(c) Use the estimated parameters in the model to obtain the forecast.

An obvious development is to try to reduce the number of steps. In the example above we have in fact done this, for we wrote down intuitively recurrence relations for the estimators $\tilde{\mu}$, $\tilde{W}$ and $\tilde{\alpha}$. We did not derive these estimators from any principle of estimation. In practice this is frequently done. It can sometimes be shown that the estimators obtained are in fact the same as would have been obtained from steps (a) and (b) above. In general, however, it is difficult to say much about the properties of such estimators. A detailed study of a certain type of recursive estimator is given by Albert and Gardner (1966).

Brenner, D'Esopo and Fowler (1968) examine the properties of sets of recurrence relations that give recursive estimators for parameters in polynomial prediction. For example, for a quadratic model they consider the set of relations

$$\tilde{\alpha}_t = ax_t \qquad\qquad + (1-a)(\tilde{\alpha}_{t-1} + \tilde{\beta}_{t-1})$$

$$\tilde{\beta}_t = b(\tilde{\alpha}_t - \tilde{\alpha}_{t-1}) + (1-b)(\tilde{\beta}_{t-1} + \tilde{\gamma}_{t-1})$$

$$\tilde{\gamma}_t = c(\tilde{\beta}_t - \tilde{\beta}_{t-1}) + (1-c)\tilde{\gamma}_{t-1}$$

with the forecasting formula

$$\tilde{x}_{t,h} = \tilde{\alpha}_t + h\tilde{\beta}_t + \frac{h(h+1)}{1.2}\tilde{\gamma}_t$$

A further step is obviously to try to go directly to step (c), or rather to go directly to a forecasting formula without reference to model parameters at all. One approach is to specify the form of forecasting formula and study its properties. For example, Ward (1963) shows that the recurrence relation

$$\tilde{x}_t = \alpha x_t + \beta x_{t-1} + (2-\alpha)\tilde{x}_{t-1} - (1+\beta)\tilde{x}_{t-2}$$

includes several standard trend predicting formulae as special cases. He then examines some of the properties of forecasts obtained by such a formula. Yet another approach is to start with a very general forecasting formula and choose its parameters to get the best forecasts irrespective of the underlying model. We consider in section 14.5 the most common way of doing this.

14.4 Extensions to the error correction method — Kalman models*

We showed in Chapter 5 that the linear trend model could be forecast using recurrence relations for the estimated current mean μ_t and slope β_t. We also noted that these relations could be expressed in an error correction form, i.e.

$$\tilde{x}_{t+1} = \hat{\mu}_t + \hat{\beta}_t$$

$$\hat{\mu}_t = \hat{\mu}_{t-1} + \hat{\beta}_{t-1} + (1-a)e_t$$

$$\hat{\beta}_t = \hat{\beta}_{t-1} + ce_t$$

where e_t is the one-step-ahead forecast error and a and c are forecasting parameters. In section 7.6(b) (iii) we noted that these results could be obtained by using a recurrence formulation for the trend model itself, namely

$$x_t = \mu_t + \epsilon_t$$

$$\mu_t = \mu_{t-1} + \beta_{t-1} + \gamma_t$$

$$\beta_t = \beta_{t-1} + \delta_t$$

the quantities ϵ_t, γ_t and δ_t being independent random variables. It is found that many of the structures discussed in previous chapters can be described by models having such a recurrence form. For models of this form it can be shown that the least squares and discounted least squares estimators of the model parameters can be expressed in error correction forms, analagous to the models themselves. Having illustrated this for the linear trend model, let us give a sketch of this formulation for two other structures from previous chapters, namely for seasonal and growth curve structures.

(a) For the seasonal index models of section 9.2.2 a recursive model of the additive form would be

$$x_t = \mu_t + \phi_{t,j} + \epsilon_t$$

$$\mu_t = \mu_{t-1} + \gamma_t$$

$$\phi_{t,j} = \phi_{t-r,j} + \delta_{t,j} \ (j = 1, \ldots, r)$$

Thus the seasonal index $\phi_{t,j}$ for each season, j, is changing by a random amount at each season. The recursive formulae for estimating the mean and the seasonal indices are

$$\hat{\mu}_t = \hat{\mu}_{t-1} + ae_t$$

$$\hat{\phi}_{t,j} = \hat{\phi}_{t-r,j} + c_j e_t \ (j = 1, \ldots, r)$$

where a and c_j are appropriate constants.

(b) As a further useful example, consider a model analogous to the simple modified exponential growth curve. In the model for this, given in Chapter 9, we have a linear model if we can assume the growth rate to be known. An analogous model here is to introduce an assumed known growth term into the linear trend model. Thus we rewrite the linear trend model as

$$x_t = \mu_t + \epsilon_t$$

$$\mu_t = \mu_{t-1} + k\beta_{t-1} + \gamma_t$$

$$\beta_t = k\beta_{t-1} + \delta_t$$

where k $(0 < k < 1)$ is the parameter that controls the growth. Here the expected value of β_t tends to zero and the curve tends to an asymptotic value. The corresponding forecast and estimators of error correction form are thus

$$\hat{x}_{t+1} = \hat{\mu}_t + \hat{\beta}_t$$
$$\hat{\mu}_t = \hat{\mu}_{t-1} + k\hat{\beta}_{t-1} + A_1 e_t$$
$$\hat{\beta}_t = k\hat{\beta}_{t-1} + A_2 e_t$$

As an aside, it should be noted that in practice k will usually be an unknown parameter. This may be intuitively estimated from the data, for example, by exponentially smoothing the quantities

$$z_t = (x_t - \hat{\mu}_{t-1})/\hat{\beta}_{t-1}$$

We have now seen three examples of linear models based on recurrence relations with analagous error correction formulae for updating the estimates of the parameters. The theory and application of linear models in recurrence form was developed by Kalman and Bucy (1961) in the early 1960s and further developed for general forecasting use in the early 1970s by Harrison and Stevens (1971). The basic form of the Kalman results are:

(a) x_t = a linear function of a set of time-dependent parameters θ_t + a random variable ϵ_t with zero mean and constant variance
e.g.

$$x_t = 1 \times \mu_t + 0 \times \beta_t + \epsilon_t$$

In matrix form we write, in general,

$$x_t = V\theta_t + \epsilon_t$$

V being a known quantity.

(b) A set of equations that relate the current values of each of the parameters to a linear function of past values of all the parameters, with added random variation. e.g.

$$\mu_t = 1 \times \mu_{t-1} + 1 \times \beta_{t-1} + \gamma_t$$

$$\beta_t = 0 \times \mu_{t-1} + 1 \times \beta_{t-1} + \delta_t$$

The general form is

$$\theta_t = L\theta_{t-1} + MY_t$$

L, M being known quantities and Y_t representing the random variables.

(c) For the above models it can be shown that the forecast having the smallest mean square error is given by

$$\hat{x}_{t+1} = VL\hat{\theta}_t$$

where $\hat{\theta}_t$ is obtained by an error correction formula of the form:

$$\hat{\theta}_t = L\hat{\theta}_{t-1} + K_t e_t$$

The form for $\hat{x}_{t+1}$ follows naturally from the equations of the model. In the error correction formula $L\hat{\theta}_{t-1}$ gives the natural updating of the parameters and $K_t e_t$ provides a correction based on the latest forecast error e_t. The correction multiplier K_t is best discussed in relation to three possible situations that can arise:

(i) If we have a deterministic model or a stochastic model with fixed parameters, treated in a global fashion, the unknown parameters do not change in a random fashion. Thus $\gamma = 0$. In this situation we would expect to get better and better estimators of the parameters θ as time goes by. Thus we will take less and less notice of the errors e_t. For this to happen K_t will approach zero as t increases. We have already seen an example of this in Chapter 4 where, for our global constant mean model, we can write

$$x_t = \mu_t + \epsilon_t$$

$$\mu_t = \mu_{t-1}$$

$$\hat{\mu}_t = \hat{\mu}_{t-1} + \frac{1}{t}\,e_t$$

Thus the correction multiplier is $K_t = 1/t$ which decreases as t increases.

(ii) If the values of the parameters θ are influenced by the occurrence of the random variables γ, then there is no fixed sequence of values to which $\hat{\theta}_t$ can move. Thus K_t will not necessarily go to zero. In this situation K_t will depend on the variance properties of all the random components. It is found that the values of K_t, in both this case and in (i) above, can themselves be found using recurrence relations. The detailed derivation and use of these is beyond the scope of this book and the reader is referred to Harrison and Stevens (1971) and Kalman and Bucy (1961) for such detail.

(iii) In (ii) the assumption of constant θ, used in (i), was avoided by defining a precise model for θ_t. The multiplier K_t requires detailed knowledge of the variance properties of the process for its evaluation. As an alternative approach we may regard the situation of (i) as a model to be used in a local fashion. We could use discounted least squares to derive the recurrence relations for K_t (see, for example, Morrison, 1969). Alternatively, we could simply regard K_t as a forecasting parameter, as indeed we did in our study of Holt's method in Chapter 5.

In this case K_t can be chosen simply as a fixed, or possibly slowly changing, quantity whose value is chosen to give the best forecasts.

(d) The ideas of 'Bayesian' forecasting introduced by Harrison and Stevens (1971) extend the approach indicated in (a), (b) and (c) by allowing the random variable γ to be in a number of possible states. The underlying concepts of this extension were illustrated at the end of section 10.3.

14.5 Linear forecasting formulae

A considerable number of the forecasting formulae that we have studied in the previous chapters are linear in the past observations. That is to say, they take the form

$$\tilde{x}_t = b_1 x_t + b_2 x_{t-1} + \ldots + b_r x_{t-r+1}$$

For example:

(a) The forecast based on a simple moving average of r terms is

$$\tilde{x}_t = \frac{1}{r} x_t + \frac{1}{r} x_{t-1} + \ldots + \frac{1}{r} x_{t-r+1}$$

(b) A forecast based on exponential smoothing, using the approximate form for large r, is

$$\tilde{x}_t = (1-a)x_t + (1-a)ax_{t-1} + \ldots + (1-a)a^{r-1} x_{t-r+1}$$

(c) The forecast for the autoregressive model of order r is

$$\tilde{x}_t = \phi_1 x_t + \phi_2 x_{t-1} + \ldots + \phi_r x_{t-r+1}$$

(d) It can be shown that the forecast for any moving average model or mixed autoregressive—moving average model can be expressed as an infinite series of past observations. However, in practice it is often found that the coefficients can be set at zero after some suitably chosen number of values, with very little loss in terms of forecasting accuracy. The forecast thus takes the linear form above.

We have always justified the use of such formulae on the basis of some appropriate model. We may, however, take a different approach and forget all about model building and simply find the values of the coefficients, $b_1, b_2, \ldots, b_r$, for which the formula gives the best forecasts. Such an approach has a number of weaknesses. It tells us nothing about the structure of the situation, for it does not require such knowledge to make it work. It will not necessarily produce the best forecast possible, for many structures lead to forecasts that are not simply linear functions in the observations. The best linear forecast, however, may be the overall optimum. Even if it is not, it gives a forecast which involves minimal knowledge and assumptions about the underlying situation. Thus for a purely empirical approach to forecasting there is something to be said for trying out such linear forecasting formulae. The aim of this section is to show briefly several approaches that have been used to find the values of the coefficients, $b_1, b_2, \ldots, b_t$, which provide the best forecasts.

(a) To illustrate the first method, consider the simplest linear forecast which uses only the last observation x_t to provide a forecast of x_{t+k}:

$$\tilde{x}_{t,k} = bx_t$$

As the criteria for the 'best' forecast we will use the minimum mean square error. Thus we wish to minimize

$$S = E(e_{t,k}^2) = E(x_{t+k} - bx_t)^2$$
$$= E(x_{t+k}^2) - 2\,bE(x_{t+k}x_t) + b^2\,E(x_t^2)$$

If we assume, without loss of generality, that $E(x) = 0$ and $\mathrm{Var}(x) = \sigma^2$ for all time, and denote the autocovariance of lag k by $C(k)$, i.e. $E(x_{t+k}x_t) = C(k)$, then

$$S = \sigma^2 - 2\,bC(k) + b^2\,\sigma^2$$

This expression is minimum when

$$b = C(k)/\sigma^2$$

Thus the forecasting parameter depends on k, the lead time of the forecast. To turn this into a practical method we need to estimate $C(k)$ and σ^2 from the data. The theoretical minimum value of S achieved by using this value of b is

$$S_{\min} = \sigma^2 - C(k)^2/\sigma^2$$
$$= \sigma^2(1 - \rho(k)^2)$$

where $\rho(k)$ is the autocorrelation coefficient of lag k. It will be seen from this that as the autocovariance increases the mean square error is reduced.

If we take the general linear forecasting formula and carry out the same calculations as in the example above, we obtain a set of r simultaneous equations, the normal equations, for the unknown parameters $b, \ldots, b_r$:

$$b_r C(0) + b_{r-1} C(1) + \ldots + b_1 C(r-1) = C(r+k-1)$$
$$b_r C(1) + b_{r-1} C(0) + \ldots + b_1 C(r-2) = C(r+k-2)$$
$$\ldots \qquad \ldots \ldots \qquad \ldots$$
$$b_r C(r-1) + b_{r-1} C(r-2) + \ldots + b_1 C(0) = C(k)$$

This assumes, as in the example, that $E(x) = 0$ over time, and that the autocovariances $C(i)$ depend only on the time intervals i between the x values and not on the actual time. The study of the above equations for practical use is a major subject, see Wiener (1949), Whittle (1965) and Yaglom (1962), and we limit ourselves to making the following points:

(i) Given knowledge of $C(0)$, $C(1)$, etc., there are many computer programs that can solve such sets of equations. The symmetry of the equations helps to make this a relatively fast process unless r is large.

(ii) In practice we must estimate the autocovariances. For large values of r this cannot be done very precisely. If an examination of the estimates of $C(0)$, $C(1)$, etc., indicates that they have some structure, e.g. $C(k) = \rho^k$, as for one of the models discussed in Chapter 7, then this can be used and the problem greatly simplified. It is, however, a move away from a purely empirical approach.

(iii) Robinson (1967) suggests ways of solving the equations by using first only one past observation, $r = 1$, as in our example, then two, three, etc. These methods are not only efficient but may indicate when we have sufficient past values to obtain reasonable forecasts.

(iv) If $E(x_t) = \mu$, for all time, instead of zero, we simply subtract μ (or our estimate of it) from every x. The forecast of x_{t+h} will then be our linear forecast obtained from the adjusted observations plus μ, i.e.

$$\text{Forecast} = \mu + b_1 (x_1 - \mu) + \ldots + b_t (x_t - \mu)$$

$$= k\mu + \tilde{x}_t$$

where $\tilde{x}_t$ is the linear forecast using unadjusted observations and

$$k = 1 - b_1 - \ldots - b_t$$

(v) The above method and normal equations generalize in a natural fashion to deal with the analogous multivariate forecasting problem (see Robinson, 1967).

(b) An alternative approach to this problem follows from the simple observation that the form of the forecast corresponds to a regression on past observations. The linear forecasting formula gives the fitted value $\tilde{x}_{t,k}$ corresponding to the observed value x_{t+k}. The regression coefficients $b_1, b_2, \ldots, b_r$ (plus a constant term b_0) can be chosen by least squares or discounted least squares using the normal equations of Chapter 6. This enables one to make use of standard regression programmes that use stepwise regression procedures to select the significantly non-zero values of b to use. We would put the values x_{t+k} as the dependent variable and x_t, $x_{t-1}, \ldots, x_{t-r+1}$ as the regressor variables. This approach is simpler than that of (a) and has proved to be effective in practice, (e.g. Newbold and Granger, 1974). A variation on this, discussed by Wheelwright and Makridakis (1973), uses a recursive method to find the b values instead of the normal equations.

14.6 Using mixed methods

The aim of this section is to describe briefly a forecasting situation which has a complicated structure and to show how several different methods of forecasting can be brought together in such a situation. The example chosen to illustrate this type of situation is one where it is required to forecast the number of children in different age groups in State schools in a given school area. Such forecasts are usually required for the next five to ten years. The first task in solving any such problem is the listing of the relevant factors. Clearly in this situation we must at least take account of the following factors:

(a) the number of children in the various age groups in the area at present;
(b) the number of births over the past few years, and estimated for future years, in the area;
(c) the number of new children expected to move into the area from outside during the forecast period;
(d) the number of children expected to leave the area over this period.

To these factors we have to add the further complication (e) that not all the children will go to the State schools in the area, a proportion in England, for example, will go to Roman Catholic and other private schools. The situation is summarized in Figure 14.2. To obtain the final forecast, it is thus necessary to look at the factors and see how they may each be forecast or measured. Let us look at the factors in turn:

(a) The number of children who are already in schools in the area can be used to form the basis of the forecast. For example, if the number of sever-year-old children at the start of one school year is 220 in the area, then the number of eight-year-olds at the start of the following year will be 220 plus the number of children arriving less the number of children leaving. This, however, will not work for the classes of children who start school for the first time during the period being forecast.
(b) The forecasting of these last numbers is based on the number of births in the area. If children start school at five, then the number of births in the area in past years will provide the basic information for forecasting the number of new entrants to the schools for the

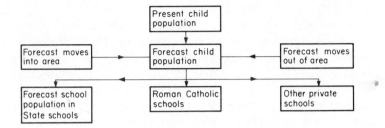

Figure 14.2 A structure for forecasting child populations

next 4 or 5 years. This would probably be 4 years, as a year will usually be lost in getting the accurate information. If we wish to forecast further into the future than this, we will have to forecast future births in the area. This can be done at the simplest level by taking the series of past births in the area and forecasting from this, using the methods developed for constant mean model, trending models or whatever seems appropriate. A more complicated approach would be to seek to form a model based on the factors thought to influence the birth rate. The simplest way to do this is to try a simple linear regression model. In one study some fifteen factors were listed as likely to influence the number of births. These included such things as the area population, number of new houses built in the year, previous birth rates and the number of women in various age groups. The final regression equation used was

Number of births in year N = 0.006 (population)

$+ 0.055$ (new houses completed in year N)

$+ 0.515$ (births in year $N - 1$)

$+ 0.227$ (Roman Catholic baptisms in year $N - 2$)

Clearly, such an equation is an empirical attempt to select the main factors in the situation and is subject to the various cautions given in Chapter 6. However, in the above situation it 'explained' 80 per cent of the variation in past data.

(c) In attempting to estimate the number of children arriving in and leaving the area a major factor is the construction of new housing and the destruction of old housing. This type of information is usually available as part of local planning information. The problem is that of relating the new and old housing types to the number of children of the various age groups involved. The procedure here is to classify the houses according to type, e.g. by the number of bedrooms and cost range. A survey of new and old properties of the various types is then carried out to obtain information on the number and ages of the children in each type. This enables the housing plans to be converted to forecast the number of incoming and outgoing children. In one study the above procedure was elaborated by allowing for the fact that supply and demand relationships will affect the pattern of occupancy of new housing. In dealing with the number and age structure of those leaving the area and those moving internally, information can be obtained via surveys and also from the records of schools in the area over past years.

(d) In a number of surveys regression methods have been used to forecast future numbers of Roman Catholic baptisms. As an example, denoting the number of baptisms in year N by B_N and the number of births by b_N, then one model used was

$$B_N = 0.20\, b_N + 0.230\, B_{N-1} + 0.32\, B_{N-2}$$

In estimating the total child population in Roman Catholic schools in one area the above relation was used for forecasting baptisms of children in the area. However, numbers of children came to local Roman Catholic schools from other areas around. Rather than seek the detailed information required to produce a regression type model for each of these areas, exponential smoothing was used on the number of such children attending in past years as obtained from school records. Thus, though a better method was available for forecasting, considerations of economy led to a less sophisticated method being used. The loss in quality of forecasts was here not so important, as the factor concerned was only one of many. A forecast of the number of children going to private schools can be obtained by studying past numbers and proportions attending such schools and using standard forecast methods on the sequences obtained.

The structure of the situation, as indicated in Figure 14.2, now enables all the various statistics and forecasts to be brought together and welded into a total model of the situation which provides the final forecasts.

References

Albert, A. E. and Gardner, L. A. (1966). *Stochastic Approximation and Non-Linear Regression*. MIT Press, Cambridge, Mass.

Bamber, D. J. (1969). A versatile family of forecasting systems. *Op. Res. Quarterly*, 20, 111–121.

Berrisford, H. G. (1965). Relation between gas demand and temperature. *Op. Res. Quarterly*, 16 229–246.

Brenner, J. L., D'Esopo, D. A. and Fowler, A. G. (1968). Difference equations in forecasting formulae. *Man. Sci.*, 15(3), 141—159.

Chen, G. Y. C. and Winters, P. R. (1966). Forecasting peak demand for an electric utility with a hybrid exponential model. *Man. Sci.*, 12(3), 531—537.

Chow, W. M. (1965). Adaptive control of the exponential smoothing constant. *J. of Ind. Eng.*, 16, 314—317.

Eilon, S. and Elmaleh, J. (1970). Adaptive limits in inventory control. *Man. Sci.*, 16(B), 533—548.

Harrison, P. J. and Stevens, C. F. (1971). A Bayesian approach to short term forecasting. *Op. Res. Quarterly*, 22, 341—362.

Kalman, R. E. and Bucy, R. S. (1961). New Results in linear filtering and prediction Theory. *J. of Basic Eng.*, 83D, 95—108.

Montgomery, D. C. (1970). Adaptive control of exponential smoothing parameters by evolutionary operation. *AIIE Trans.*, 2, 268—269.

Morrison, N. (1969). *Introduction to Sequential Smoothing and Prediction*. McGraw-Hill, New York.

Newbold, P. and Granger, C. W. J. (1974). Experience with forecasting univariate time series and the combination of forecasts. *J. Roy. Statist. Soc. A*, 137, 131—146.

Rao, A. G. and Shapiro, A. (1970). Adaptive smoothing using evolutionary spectra. *Man. Sci.*, 17, 208—218.

Roberts, S. D. and Reed, R. (1969). The development of a self-adaptive forecasting technique. *AIIE Trans.*, 1, 314—322.

Roberts, S. D. and Whybark, D. C. (1973). Adaptive forecasting techniques. *J. of Ind. Eng.*.

Robinson, E. (1967). *Multichannel Time Series Analysis with Digital Computer Programs*. Holden-Day Inc., San-Francisco.

Trigg, D. W. (1964). Monitoring a forecasting system. *Op. Res. Quarterly*, 15, 271—274.

Trigg, D. W. and Leach, A. G. (1967). Exponential smoothing with an adaptive response rate. *Op. Res. Quarterly*, 18, 53—59.

Van Dobbin de Bruyn, C. S. (1964). Prediction by Progressive correction. *JRSS (B)*, 26, 113—122.

Ward, D. H. (1963). Comparison of different systems of exponentially weighted prediction. *The Statistician*, 13, 173—185.

Wheelwright, S. C. and Makridakis, S. (1973). *Forecasting Methods for Management*. John Wiley and Sons, New York.

Whittle, P. (1965). *Prediction and Regulation by Linear Least Squares*. EUP, London.

Wiener, N. (1949). *Extrapolation, Interpolation and Smoothing of Stationary Time Series*. John Wiley and Sons, New York.

Yaglom, A. N. (1962). *Theory of Stationary Random Functions*. Prentice-Hall. Englewood Cliffs. New Jersey.

Chapter 15

The analysis and comparison of methods

15.1 Introduction

It would make life much easier if we could say that the forecasting formula on page X is the best to use and all the others are of historical interest. Unfortunately we cannot. One might say that surely knowing the model the theoretician can derive the best forecasting formula. It is true that he can, theoretically, derive the best forecasting formula given some criterion, but in practice there are many snags with this. The model we use will probably not be exactly right and even if it was good enough for the past it may not be too good now. The model may only describe certain aspects of the actual situation; other aspects will shows up in the forecasting errors. Even if all is well we need to have a forecasting formula that responds well to minor changes that might occur in the structure of the data. In general, then, we can rarely choose, for a practical situation, the appropriate method without a great deal of analysis of the data and a comparison of various possible methods. The aim of this chapter is to discuss ways of carrying out this analysis and comparison.

It is possible to approach the comparison of methods in a purely mechanical fashion; we simply calculate the mean square error for a set of methods and/or a set of values of the forecasting parameters and choose the best. However, for effective forecasting the objectives are to seek the best methods *and* to understand their properties. It might be, for example, that a more careful study of the results of the trials would have led not to choosing that method with the smallest mean square error but to developing an entirely different method. At the heart of our study is the forecast error. Most, though not all, of the methods developed in this chapter will be aimed at analysing the forecast errors to see if they show any structure of their own. If the forecast errors do show a clear structure then, as we will see in Chapter 17, we may use this knowledge to obtain better forecasts or make better use of the forecasts we have.

One or two general observations should be made in relation to the use of errors in this chapter. Firstly, one must obviously study the errors for the type of lead time of practical interest. There is little point and possible dangers in studying one-month-ahead errors if the lead time in practice is going to be three months. If there are several lead times of interest they should all be analysed, though if this is not possible some appropriate middle value should be used. Secondly, in plotting graphs and calculating statistics it should be remembered that the first few forecasts are very unlikely to be as good as the later ones. These should therefore be discarded or somehow weighted before carrying out the error analysis. Thirdly, if forecasts depend on estimates of model parameters and/or choices of forecasting parameters made from a given set of data, then it is unwise to carry out the forecast analysis on that same set of data. Having said this one must admit that frequently data is so scarce that this has to be done. The danger, of course, is that the parameters having been chosen using the data lead to a fitted model or forecasting formula that fits very snugly to that particular set of data; however, when it is applied to future data the fit will probably not be so good and the conclusions of the forecast analysis may turn out to be wrong. With sufficient past data one can avoid this problem by choosing the parameters on the basis of the first half of the data and carrying out the forecast analysis on the second half. If there is not much data, some improvement can be obtained by carrying this out as stated but then interchanging the two halves of the data and repeating the process. This process is called crossvalidation. An alternative procedure is based on the fact that many models are essentially unaltered in form by reversing the time axis. In such a case we might choose the best forecasting parameter by using the data as though x_t were the first observation, x_{t-1} the second, and so on. The forecast analysis would then be carried out using the data in the correct time sequence. When comparing different methods, there is something to be said for comparing the best possible results obtainable for each method. These we will get by using the same data for the fitting exercise as for the comparison. As long as it is realized that the results will be better than we could expect in practice, then we are on reasonably safe ground.

It will be assumed in the following sections that we have decided on an appropriate model and hence have some reasonable methods to try out. If our choice was poor the results of the analyses to be described should tell us of the fact and guide us towards more appropriate choices.

15.2 Forecast analysis

The human eye has a great ability at detecting pattern, admittedly sometimes erroneously. However, as we are looking for structure, for

222

pattern, in our data and in our forecast errors, we would be most
unwise to ignore this ability. The aim of this section is to list a number
of useful ways in which graphical and statistical analysis can help in
forecasting. We will assume that for the method of forecasting under
investigation we have a set of forecasts together with the observed
values that occurred and the consequent errors, denoted by $(\hat{x}_i, x_i, e_i)$,
$i = 1, \ldots, t$. The main types of procedure of practical use are as
follows.

15.2.1 Plots of forecasts and actual observations $(\hat{x}_i, x_i)$

(a) Double plot against time

In Figure 15.1 we have simply plotted forecasts and observations on
the same graph. This gives an immediate impression of how well the
forecasts are following the observed data.

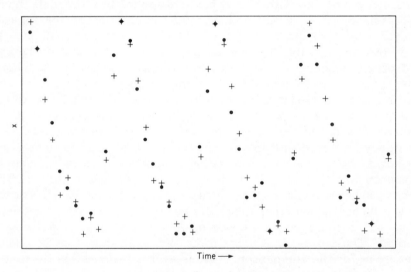

Figure 15.1   Plot of data ($\cdot$) and forecast ($+$) superimposed

(b) Forecast—observation diagram

In Figures 15.2(a) the forecasts are plotted against the observations. If
all the forecasts were perfect, all points would lie on the $45°$ line shown
in Figure 15.2(a), termed the line of perfect forecasts. In the figure
there is a tendency for the points to lie below this line, thus indicating a
bias in the forecasts causing them to be larger on average than the
observed values.

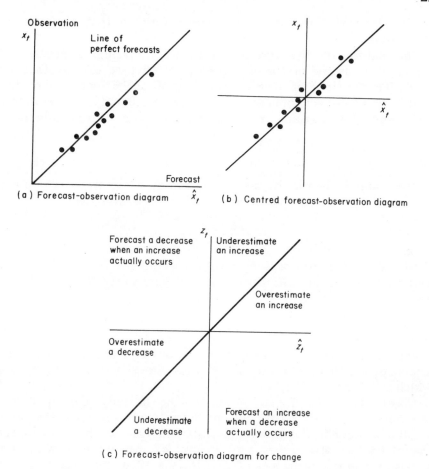

(a) Forecast-observation diagram   (b) Centred forecast-observation diagram

(c) Forecast-observation diagram for change

Figure 15.2   Forecast—observation diagrams

(c) Centred forecast—observation diagram

In Figure 15.2(b) we have moved the axes of the forecast—observation diagram so that they are somewhere near the middle of the data. The numerical value at the intersection point must be the same for both $\hat{x}$ and x, but its exact value does not matter too much as long as there is a reasonable spread of data to either side of the axes. Again we have a line of perfect forecasts at 45° to the axes. The actual points shown on Figure 15.2(b) show a tendency to fall below the line of perfect forecasts in the upper right-hand quadrant and an opposite tendency in the lower left-hand quadrant. Thus for large observations there is a tendency to overforecast and for the smaller observations there is a tendency to underforecast. An effect of this would be to give nearly unbiased forecasts with a larger variance than the original observations.

The nature of the forecast—observation diagram suggests two numerical measures of forecast quality. Firstly, we can say that we would like the slope of the regression line of x on $\hat{x}$ to be close to that of the line of perfect forecasts, i.e. a slope of 1.0 and an angle of 45°. The formula for this slope, from the ordinary theory of regression, is

$$\text{Slope} = m = \frac{\Sigma(x_i - \bar{x})(\hat{x}_i - \bar{\hat{x}})}{\Sigma(\hat{x}_i - \bar{\hat{x}})^2}$$

where $\bar{x}$ and $\bar{\hat{x}}$ are the averages of x_i and $\hat{x}_i$, respectively. The other measure is the correlation between x and $\hat{x}$; this is given by

$$r(x, \hat{x}) = r = \frac{\Sigma(x_i - \bar{x})(\hat{x}_i - \bar{\hat{x}})}{\sqrt{[\Sigma(x_i - \bar{x})^2 \, \Sigma(\hat{x}_i - \bar{\hat{x}})^2]}}$$

If we had perfect forecasts, $\hat{x}_i = x_i$, this would give $r(x, \hat{x}) = 1$. Thus a large value of r is an indication of a good forecast. Notice the relation between m and r, namely

$$m = r \frac{s_x}{s_{\hat{x}}}$$

where s_x and $s_{\hat{x}}$ are the standard deviations of x and $\hat{x}$, respectively. The line of perfect forecasts would thus correspond to forecasts for which $m = 1$, $r = 1$ and $s_x = s_{\hat{x}}$.

Clearly the values of r and m are of value for studying the quality of our forecasting methods for all our data. It often happens that local variations occur in the underlying situation. This has the consequence that the values of r and m vary locally. We may estimate such local values by taking moving sets of data for the calculations, as in moving averages. Alternatively, we can introduce a weighting factor in the summations for which the weights die away in both directions. This is most easily done by using weights that correspond to exponential smoothing. This is illustrated in Appendix C. Figure 15.3 shows local correlation, $r(x_t, \hat{x}_t)$, derived in this fashion, plotted against t.

(d) Forecast—observation diagrams for change

In some situations one is interested in a method's ability to forecast increases or decreases in a variable, e.g. whether sales are going to go up or down next month. Thus we are interested in changes from the present sales x_t. Thus let

$$z_t = x_{t+1} - x_t$$

denote the observed change in sales and

$$\hat{z}_t = \hat{x}_{t+1} - x_t$$

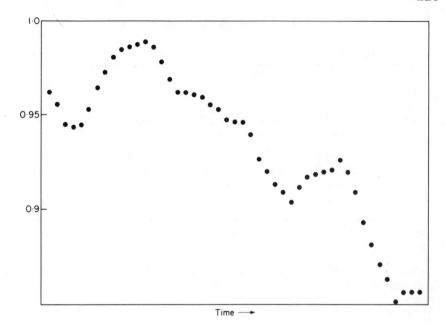

Figure 15.3   Plot of local correlation

denote the forecast change in sales. The quantities $\hat{z}_t$ and z_t can now be plotted on a forecast—observation diagram as in (c) above. The interpretation of the points obtained are as indicated in Figure 15.2(c). If points lie in the upper left or lower right quadrants, the inference is that a 'turning point' error has been made; $\hat{z}$ and z have opposite signs. In the other two quadrants the errors are in terms of magnitude of change but not of direction. A detailed discussion of this type of plot, together with many examples, is given by Theil (1958).

15.2.2 Analysis based on forecast errors

The plots and statistics here are all based on studying the forecast errors. In the discussion we will use the ordinary error $e_t = x_t - \hat{x}_t$. It is, however, sometimes useful to use the percentage error $100\ e_t/x_t$ in place of e_t in the analyses.

(a) Tabulating forecast errors

The most obvious analysis of the errors is to simply look at the errors themselves. Table 15.1(a) shows a small part of such a detailed analysis. The errors given are those made in forecasting the observation made at the given time from h time units previously, $h = 0$ corresponding to simple smoothing.

Table 15.1 Tabulating forecast errors

(a) Part of a list of errors

Time	0	1	2	3	4
			Lead time h		
20	−0.004	0.066	0.236	0.407	0.377
21	0.016	0.170	0.324	0.278	0.232
22	0.042	0.154	0.065	−0.024	0.088
23	0.038	−0.089	−0.216	−0.143	−0.070
24	−0.022	−0.127	−0.032	0.063	0.258

(b) A set of standardized errors

Standardized errors	0.81	1.46	0.84	−0.69	−0.45	0.10	−0.23	−1.92
Ordered standardized errors	−1.92	−0.69	−0.45	−0.23	0.10	0.81	0.84	1.46

Another useful device is to standardize the errors, i.e. subtract $\bar{e}$ and divide by their standard deviation s_e. If the data are approximately normal, 95 per cent of these should lie between −2 and +2. Anything much greater than 3 can be regarded as an outlier. Table 15.1(b) shows a set of such standardized errors in their order of occurrence and in their order of magnitude.

(b) Plot of error against time

In Figure 15.4(a) the simplest plot of e_t against time is shown. This gives the clearest indication of the nature of the forecast errors. It will indicate their magnitude, roughly how independent adjacent errors are, the existence of trends or other systematic changes in the errors and the occurrence of any particularly large errors.

(c) Histogram of forecast errors

The first question about the statistical properties of errors concerns their distribution. The histogram or bar chart such as that in Figure 15.4(b) provides one of the easiest presentations of the sample distribution. The area under each bar is equal to the frequency of errors occurring in the corresponding range of values of e. If equal intervals of e are used, as in the example (b), the vertical height can be used to represent frequency.

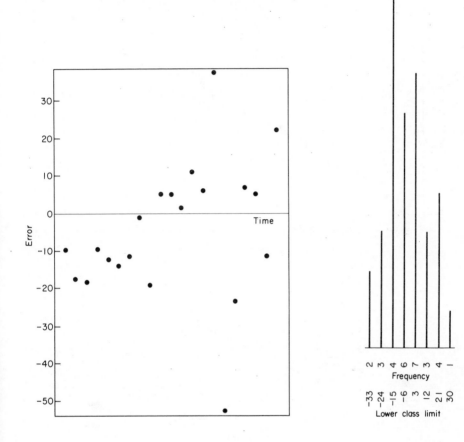

Figure 15.4   Plots of forecast errors

(d) Normal plots

Suppose the histogram looks reasonably symmetrical and it is desired to construct some sort of limits to forecasts based on assuming normally distributed errors. In this situation it is useful to carry out a further check on the normality of the error distribution. This can be done with the use of normal plots, such as that illustrated in Figure 15.5(a). This may make use of commercially available graph paper. The axes of this paper are so designed that plots of the data are approximately straight lines if the data come from a normal population; otherwise it reveals some form of twisting. Figure 15.5(b) indicates how to interpret the standard normal probability plot.

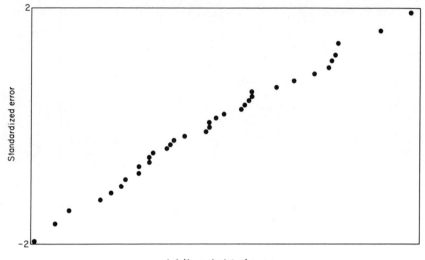

(a) Normal plot of errors

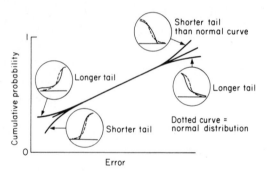

(b) Interpretation of deviation from straight line

Figure 15.5   Normal plots

(e) Errors against forecasts

This plot, illustrated in Figure 15.6, gives an indication as to whether there is any relation between the size and sign of the errors and the magnitude of the forecasts. If we regard the forecast as a rough estimate of the underlying mean, then this plot gives an indication as to whether the magnitude or sign of the forecast errors are effected by the magnitude of the underlying mean. An immediate application of this occurs in the distinction between the additive and the multiplicative occurrence of the irregular component in the seasonal model. In the additive form the errors are not influenced by the size of the seasonal. If, however, we have a situation which is multiplicative, then the

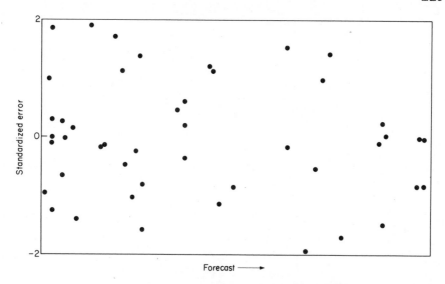

Figure 15.6   Plot of error against forecast

forecasting errors will tend to be greater when the seasonal peaks occur. This will show up as a greater spread in the errors for large $\hat{x}$.

(f) The distributions of x and $\hat{x}$

The plots that we have used to look at forecast errors can obviously be used to look at the distributions of the observations and of the forecasts. If the means of x and $\hat{x}$ are following some deterministic curve, a trend or a seasonal for example, such plots will not be very meaningful. However, even if we do have such a situation, it is still informative to study the conditional distribution of x_{t+h}, given $\hat{x}_{t+h}$. Thus we are concerned with finding the distribution of x_{t+h} knowing the value of our forecast $\hat{x}_{t+h}$, e.g. with the distribution of the future sales relative to our current forecast of their value. We denote this distribution by $f(x \mid \hat{x})$. Referring to the forecast—observation diagram (see Figures 15.2 and 15.7), this distribution is the distribution of the observation x lying in the line of fixed $\hat{x}$.

Without a vast amount of data or an assumption of normality this is very difficult to obtain directly. We can, however, proceed in a different fashion. It can be argued both theoretically and intuitively that the forecast errors in many situations should be independent of the magnitudes of forecasts. Plots such as Figure 15.6 give an immediate visual check on this. Our distributional studies of the forecast errors in (c) and (d) should give us some idea of the error distribution $f(e)$. Because of the independence assumption this is identical to the conditional distribution of the error, given $\hat{x}$, $f(e \mid \hat{x})$.

230

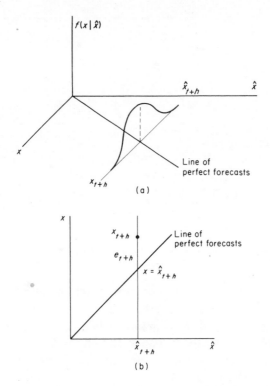

Figure 15.7 Conditional distribution of x_{t+h} given x_{t+h} (a) (b)

Writing $e = x - \hat{x}$ we now know the distribution

$$f(x - \hat{x} \mid \hat{x})$$

Thus the distribution of the future x_{t+h} is obtained by replacing e_{t+h} by $x_{t+h} - \hat{x}_{t+h}$, $\hat{x}_{t+h}$ being a known forecast, in the distribution of e_{t+h}. This is seen with reference to Figure 15.7(b). Provided that the distribution of e_{t+h} does not depend on $\hat{x}_{t+h}$ the distribution of x_{t+h} about $\hat{x}_{t+h}$, on the line of perfect forecasts, is simply the distribution of e_{t+h}. The simplest example is the case where e_{t+h} is normally distributed, $N(0,\sigma_e^2)$. It follows that x_{t+h} is also normally distributed, $N(\hat{x}_{t+h}, \sigma_e^2)$.

(g) Comparison of errors from different methods

Instead of comparing just the mean square errors of different methods it is often useful to compare the individual errors. This is done, as in Figure 15.8, by simply plotting the errors for both methods against time or on a scatter diagram.

231

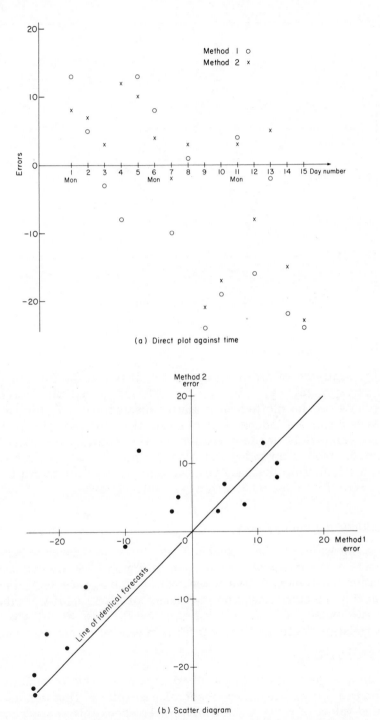

(a) Direct plot against time

(b) Scatter diagram

Figure 15.8   Comparison of errors from different methods

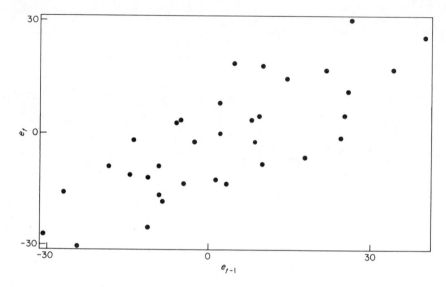

Figure 15.9   Autoregression of errors

(h) Autoregression of errors

If the sequence of errors $e_1, e_2, \ldots, e_t$ is put in the form of $t-1$ pairs, (e_1, e_2), $(e_2, e_3), \ldots, (e_{t-2}, e_{t-1})$, (e_{t-1}, e_t), the corresponding points can be plotted on a scatter diagram as in Figure 15.9. If successive errors are independent of each other, the plot obtained will show a completely random scatter. If the points look like those illustrated, there is an indication of a relationship between successive errors, a relationship that may be used to improve future forecasts. The general term for such a relationship is an autoregression.

(i) Autocorrelation of errors

As a measure of the extent of the linear relationship between successive errors in the scatter diagram of Figure 15.9, we can use the correlation coefficient between the pairs (e_i, e_{i+1}), $i = 1, \ldots, t-1$. We call this the autocorrelation coefficient of lag 1, r_1. If we had used pairs two apart, (e_i, e_{i+2}), $i = 1, \ldots, t-2$, this would give the autocorrelation coefficient of lag 2. Thus in general we can calculate

$$r_k = r(e_i, e_{i+k})$$

If we have a good forecasting method, it is reasonable to require that the forecast errors are independent of each other. This implies that expected values of $r_1, r_2, \ldots$ are all zero. To check this we calculate r_1, $r_2, \ldots$ and plot these as shown in Figure 15.10. Such a plot is called a

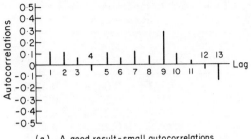

(a) A good result - small autocorrelations

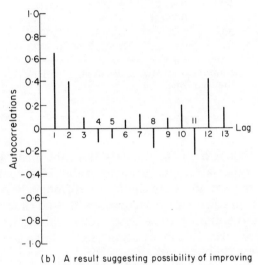

(b) A result suggesting possibility of improving
forecasts

Figure 15.10   A correlogram

correlogram. If any of the r values are significantly different from zero, the forecasting method can probably be improved. One suitable method is discussed in Chapter 17. A particular application of this plot is that of testing the success of a seasonal forecasting model. If we have failed to adequately model the seasonal components, the autocorrelation of lag corresponding to the length of the season will tend to be large.

15.2.3 Root mean square error analysis

The root mean square error is the most commonly used forecasting criterion. Thus several approaches to forecast analysis depend on a study of the root mean square error (RMSE).

Table 15.2 Comparison of root mean
square errors

Lead time	I	Method II	III
0	0.0361	0.0358	0.0286
1	0.1451	0.1449	0.1439
2	0.2144	0.2115	0.2101
3	0.2809	0.2817	0.2816
4	0.3325	0.3342	0.3347
5	0.3961	0.3981	0.3982
6	0.5009	0.5021	0.5016
7	0.6177	0.6183	0.6153
8	0.7268	0.7273	0.7237
9	0.7681	0.7696	0.7602
10	0.8822	0.8855	0.8723

Number of errors used equals 37 minus
the lead time.

(a) Tabulation

The most obvious step is to tabulate the root mean square error or the appropriate criteria for the problem. Table 15.2 compares the root mean square error for three methods for different lead times. Notice how the root mean square error increases with lead time. A natural step following this tabulation is to seek to model this increase relating the root mean square error to the lead time. In Table 15.2 a model giving a linear relation would not be far out.

(b) Plot of local root mean square error

As mentioned in section 15.2.1(c) and illustrated in Figure 15.3, we can readily calculate a local value of any statistic using the weighting technique described in Appendix C. Thus, instead of relying on a single value of the root mean square error for all our data, we can calculate its local value and plot this quantity against time. Figure 15.11 shows such a plot. We see in this example a clear indication that our forecasting method has been getting increasingly inaccurate.

(c) Plots of local bias and standard deviation

We have noted before that the mean square error can be broken into contributions from the bias and from the error variance. Thus, in theoretical, population, terms

$$E(e^2) = E(e)^2 + \text{Var}(e)$$

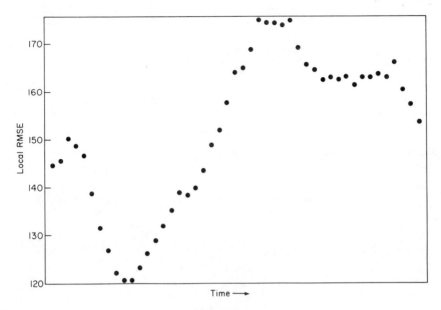

Figure 15.11   Plot of the local root mean square error

Having looked at the local root mean square error, it is natural to see how its fluctuations are related to the separate local fluctuation in the bias, $E(e)$ and the standard deviation of error, $\sqrt{\mathrm{Var}(e)}$. Figure 15.12(a) and (b) show such local plots for the same data as that used in Figure 15.11. It is clear from these that the fluctuation in the root mean square error is produced almost totally by that in the standard deviation. The bias is very small and its local fluctuations have little effect on the root mean square error. Had we been in a different situation and had large local biases occurring, then we could improve our forecasts by making some adjustment to remove the bias. We will examine this possibility in Chapter 17.

(d) Theil's coefficients

Having split the population mean square error into components, we can in fact go a stage further and split the error variance into two components. Thus

$$\mathrm{Var}(e) = \mathrm{Var}(x - \hat{x})$$
$$= \sigma_x^2 + \sigma_{\hat{x}}^2 - 2\sigma_x \sigma_{\hat{x}} \rho(x,\hat{x})$$

using a standard statistical result. Here σ_x and $\sigma_{\hat{x}}$ are the population standard deviations for observations and forecasts and $\rho(x,\hat{x})$ is the

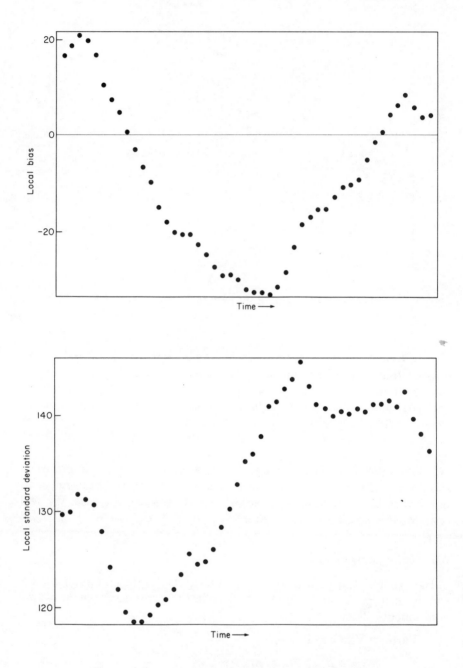

Figure 15.12   Plots of local bias and standard deviation

population correlation coefficient. Replacing these by our estimates s_x, $s_{\hat{x}}$ and $r(x,\hat{x})$, and noting that the bias from the data is just the average error, $\bar{e}$, we have

$$\text{MSE} = \bar{e}^2 + s_x^2 + s_{\hat{x}}^2 - 2s_x s_{\hat{x}} r$$

A little algebra will show that we can rewrite this as

$$\text{MSE} = \bar{e}^2 + s_{\hat{x}}^2 \left(1 - r\frac{s_x}{s_{\hat{x}}}\right)^2 + s_x^2(1 - r^2)$$

$$= \text{bias} + \text{regression} + \text{correlation}$$
$$\text{term} \quad \text{term} \quad\quad \text{term}$$

These three terms have a clear interpretation. The bias, as we have seen, indicates under- or overforecasting. The regression term, from the results of section 15.2.1(c), measures how much the regression line of x on $\hat{x}$, in the forecast—observation diagram, differs in angle from the $45°$ line of perfect forecasts. The regression term is, in fact, $s_{\hat{x}}^2 (1 - m)$, m being the slope of the regression line. The correlation term depends on r, the correlation $r(x,\hat{x})$ between observations and forecasts. If we divide all through our expression by the mean square error, the left-hand side becomes one and the positive fractions on the left-hand side will indicate the proportion of the mean square error that is due to the bias, regression and correlation aspects of the forecasts. The three terms in this form are called Theil's coefficients (see Theil, 1958 and 1965, and also Granger and Newbold, 1973). As the three coefficients add to one, they can be plotted on triangular graph paper as in Figure 15.13. The mean square error for the forecasts leading to points A_1 and A_2 is due mainly to poor correlation between forecasts and observation. For B_2 the mean square error is due largely to a bias in the forecasting. For B_1 the mean square error is due to both bias and poor correlation. For C_1 and C_2 all three factors contribute to the mean square error. Note that these interpretations depend on the easily verified assumption that s_x^2 and $s_{\hat{x}}^2$ are roughly the same.

15.3 Choosing forecasting parameters

We have so far discussed error analysis from a general investigational point of view. We turn now to the much more specific aim of the choice of the numerical values of forecasting parameters. It is assumed here that the method has been decided upon and the only task left is to choose these parameter values. The essence of the operation is to decide on the forecasting criterion to be used and then, in some fashion, to find the value of the forecasting parameter for which this achieves its best value. In the great majority of practical analyses of this type the mean square error is used as the criterion, though a weighted mean

238

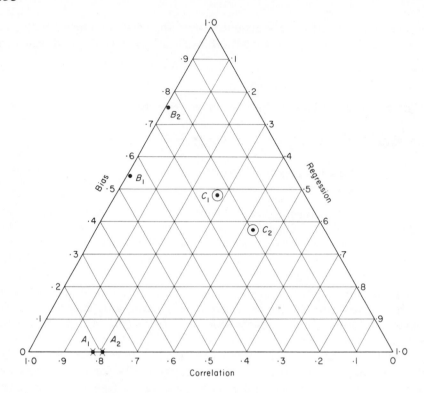

Figure 15.13   Plot of Theil's coefficients

square error may be appropriate where there is a long sequence of past data. In most of the forecasting formulae we have discussed there have been only one or two forecasting parameters. It is also clear that from the nature of these parameters we do not require very accurate specifications of them. For example, with most applications of exponential smoothing a can be chosen to within 0.05 quite satisfactorily. Hence there is usually no need to conduct any sophisticated search to find optimum values. The simplest approach is just to use past data to evaluate the mean square error at a lattice of points of possible interest and plot the result.

Table 15.3 gives a number of illustrative examples. From these a number of general points can be made:

(a) Where the forecasting parameter is the parameter a from discounted least squares then, if a_0 is the optimum,
   (i) $a_0 = 1$ corresponds to ordinary least squares and the model is being used as a global model.
   (ii) for stochastic and regression models values of a_0 close to one ($a_0 \simeq 0.9$) are the norm.

Table 15.3 Values of mean square error for various discounting factors a

Data	1.0	0.9	0.8	0.7	0.6	0.5	0.4	0.3	0.2	0.1
Set 3	46	42	39	36	33	32	31	30	31	32
Set 2	86	85	87	92	97	103	—	—	—	—
Set 1	39	40	42	45	49	55	—	—	—	—

(iii) if a_0 is small, e.g. $a_0 = 0.5$, or much greater than one, e.g. $a_0 = 1.3$, then this is taken to indicate that the model being used is inappropriate.

(iv) the curves of the criterion function may be fairly flat near a_0. This, together with the fact that a_0 is a sample quantity with its own random variation, gives the forecaster some leeway in fixing the value of a to use in practice.

(b) Where there are more than one parameter, a more elaborate lattice is needed, as, for example, with Holt's method, see Table 15.4(a). If several parameters are to be optimized for a large number of sets of data, more efficient 'hill climbing' methods should be used.

(c) The choice of model out of a class of models, e.g. of p for an

Table 15.4 Choosing two forecasting parameters

(a) Parameters a and b in Holt's method

MSE tabulated		Mean parameter a 0.5	0.6	0.7	0.8	0.9	0.99
Slope parameter b	0.5	368	352	345	351	359	370
	0.6	341	326	324	328	331	358
	0.7	326	313	305	316	319	337
	0.8	314	302	294	299	308	320
	0.9	305	293	285	289	294	390
	0.99	315	305	296	302	312	322

(b) Discounting parameter a, order p, of autoregressive model

MSE tabulated		Order p 1	2	3	4	5	6
Discounting parameter a	1.0	925	916	907	891	853	848
	0.98	890	880	869	867	814	824
	0.96	901	884	879	878	823	847

autoregressive model AR(p), is structurely equivalent to choosing a forecasting parameter, as is illustrated by Table 15.4(b).

15.4 Comparison of methods

We now turn briefly to the comparison of methods. It might seem odd to deal with this after discussing the choice of forecasting parameters, however a little thought shows that if we are to compare methods we must first choose the best parameter values for those methods. We thus compare the best results that each method can achieve. The comparison of three methods in Table 15.4, for example, in fact corresponded to a comparison of double exponential smoothing, discounted least squares trend fitting and extrapolation and Holt's method. In the first two cases the single parameter was chosen to give the smallest mean square error for a particular lead time and in the latter case the two parameters were chosen in this way. Such choices of the best method use one given criterion, but that does not mean to say that this criterion must be the only quantity used in the comparison. Indeed, if we wish to compare methods for a given type of data, it would be unwise to compare them only by error analysis on past data; some theoretical and simulation comparison should also be carried out to study the behaviour of the methods under varying circumstances. As we indicated in Chapter 3, it is very important to ask questions about the stability and robustness of one's methods; this is particularly important if the methods are to be used on a regular basis, say as part of a computer program for stock control. If one has a good set of past data some useful information on these aspects can be gained by adding, for example, step-changes or trends to the real data, forecasting these 'adulterated' data and then carrying out an error analysis. If one is using a simple linear forecasting formulae, which depend simply on some form of weighted sum of the observations, the new forecast equals the old forecast plus the forecast of the step or trend. This exercise would not then provide any more information than that obtained by a general study of the method, without reference to ones own data. However, many methods, particularly those involving seasonal aspects, are not linear and useful information about the methods being studied can be obtained by the use of 'adulterated' data. If more extensive studies are required, then it is best to simulate data from a range of situations that might foreseeably occur in the practical field being studied. These data would then be used in the various methods of forecasting and error analyses carried out.

A number of papers have appeared that compare both theoretically and practically some of the methods discussed in this book (e.g. Davis and Huitson, 1967; Godolphin and Harrison, 1975; Granger and Newbold, 1973; Groff, 1973; Marchant and Hockley, 1970; McKenzie,

1974; Newbold and Granger, 1974; Wagle, Rappoport and Downes, 1968). Rather than discuss these papers, the points made about the various methods have simply been included in the relevant sections of the book. However, the point that does emerge from this and other work is that there is certainly no best method for all situations and that what is best by one criterion may not be by another. The problem is that there are many features of a method to consider and not just its mean square error properties. One method will produce a forecast of reasonable properties as soon as three or four observations are available; another method will require sixty observations before it can be really started. One method will require only storage of the last observation for each series being forecast; another will require the last thirty observations. One method can be implemented and understood by a clerk; another will require a medium sized computer and a person with a mathematics degree to understand it. One method will perform well at short lead times but badly at long ones. One method will not be much affected by a couple of wild values; another will be badly out for a long period after the trouble. One method will involve the use of several sets of information that have value on their own and will produce useful ancillary information; other methods will just produce the forecast.

Clearly, with so many aspects to consider, each forecaster must choose his own methods to fit his own situation. A brief survey of individual practice shows that this is exactly what happens. There are nearly as many methods and approaches as there are forecasters. This sometimes produces a feeling of a 'Tower of Babel', but the above comments should indicate that it is not entirely due to human cussedness.

15.5 Prediction intervals

Suppose we are in the position of having a satisfactory forecasting method and a forecast of x_{t+h} obtained from it. What can we say about x_{t+h}? In particular, can we give an interval, a prediction interval $(x_{t+h,e}, x_{t+h,u})$ in which x_{t+h} has a high, known, probability of falling? There are two possible approaches to solving this problem: one theoretical, based on the model used, and one empirical, based on the observed behaviour of past forecast errors. Let us consider these in turn.

(a) Intervals derived from the forecasting model

We first came across the idea of confidence limits as a means of getting a prediction interval in section 4.1. In essence the discussion given there can be applied in any situation where we can use the model to find the variance of the forecast error and where that error can reasonably be taken to be normally distributed. In past chapters we have looked at a range of such formulae from

$$\text{Var}(e) = \frac{t+1}{t}\,\sigma^2$$

for the constant mean model, with the average used as a forecast, to

$$\text{Var}(e) = \text{Var}(b)x^2 + \sigma^2 + \text{Var}(\hat{x})\{\, \text{Var}(b) + \beta^2 \,\}$$

when forecasting with a regression model; in many of these cases the normality of ϵ in the model leads directly to normally distributed errors. In general, given $\text{Var}(e) = \sigma_e^2$ as the known variance of e, we can write statements such as

$$\text{Prob}(-1.96\,\sigma_e \leqslant e \leqslant 1.96\,\sigma_e) = 0.95$$

and hence to

$$\text{Prob}\,(\hat{x} - 1.96\,\sigma_e \leqslant x \leqslant \hat{x} + 1.96\,\sigma_e) = 0.95$$

Thus we have upper and lower values that on 95 per cent of occasions, when using the method, will enclose the value of x. In using the model we usually find σ_e^2 as a function of σ^2, the variance of the ϵ values in the model. This variance has to be measured in terms of the observed variance when fitting the model. In this case σ^2 is replaced by its estimate, s^2, and 1.96 has to be replaced by the value from the tables of the t-distribution with degrees of freedom as required by the estimate s^2. Where the errors are not normally distributed the problem is more complex.

Often in standard situations using global models standard computer programs derive these intervals. The basic problem here is that these calculations are usually based on assumptions of global rather than local models. Thus any deviation of the actual structure from the model will mean that the calculated prediction interval will not lead to the correct proportions of observations inside the interval. For example, in a situation described by Williams and Goodman (1971) only about 80 per cent of observations lay inside the 90 per cent prediction intervals calculated using an assumed, regression, model.

(b) Intervals derived from past forecast errors

The second approach here is to study the distribution of past forecast errors and estimate σ_e^2 by the ordinary estimate, s_e^2, of variance. If the errors are approximately normal with zero mean, then we will have for the 95 per cent prediction intervals, using n errors,

$$\text{Prob}(\hat{x} - t_{n-1,95\,\%}s_e < x < \hat{x} + t_{n-1,95\,\%}s_e) = 0.95$$

Though there are problems if the errors are not normally distributed, the basic approach is the same. In Williams and Goodman (1971), referred to above, the magnitudes of the errors, $|e|$, were found to give a good fit to a gamma distribution. This distribution was fitted to the set of errors and prediction intervals obtained. The observed proportions of forecasts, other than those used to fit the model, that fell within the intervals was satisfactorily close to the probabilities as fixed from the gamma distribution. An alternative approach, that makes no assumptions about the distribution, is based on simply ordering the errors. Suppose, for example, we were concerned only with estimating a value for e that had only a probability of 0.1 of being exceeded. In this case the largest of ten observed errors, or the second largest of twenty observed errors, etc., would provide a rough estimate based on the observed errors alone. To use estimates of this form we need to be fairly confident that the data used to obtain these estimates come from a stable error distribution and do not contain any outliers.

15.6 Sensitivity analysis

Thus far in this chapter we have talked about the study of single forecasting methods in isolation. Often our forecasts are components of a complex situation and we may well be using several different forecasts. In such situations the assessment of the methods and the study of the influence of forecasting errors on the situation become more elaborate processes. The term 'sensitivity analysis' describes the sort of exercise the forecaster must carry out in such situations. As an example of this we return to the example of the use of mixtures of methods in forecasting school populations given in section 14.6. Suppose the forecasts being used are:

$\hat{n}_t$ = forecast number of children of given age in region at beginning of year t
$\hat{l}_t$ = forecast number of leavers during year t
$\hat{a}_t$ = forecast number of arrivals during year t
$\hat{r}_t$ = forecast proportion of age in Roman Catholic schools
$\hat{p}_t$ = forecast proportion of age in other private schools
$\hat{s}_t$ = forecast number of children in State schools

Then, allowing for only these factors, we have

$$\hat{n}_t = \hat{n}_{t-1} + \hat{a}_{t-1} - \hat{l}_{t-1}$$
$$\hat{s}_t = \hat{n}_t(1 - \hat{p}_t - \hat{r}_t)$$

Clearly, even if we had sufficient past experience to be able to quote mean square errors for the various forecasts, $\hat{n}, \ldots, \hat{p}$, we would have great difficulty in evaluating the final mean square error for $\hat{s}_t$. This

Table 15.5 Sensitivity analysis

Line No.	$\hat{n}_{t-1}$	$\hat{a}_{t-1}$	$\hat{l}_{t-1}$	$\hat{n}_t$	$\hat{p}_t$	$\hat{r}_t$	$\hat{s}_t$
1	2,300	100	50	2,350	0.05	0.05	2,115
2	2,500	200	30	2,670	0.05	0.05	2,403
3	2,300	100	50	2,350	0.04	0.04	2,162
4	2,500	200	30	2,670	0.04	0.04	2,456
5	2,300	500	50	2,750	0.05	0.05	2,475
6	2,500	500	30	2,970	0.04	0.04	2,732

could be done with some effort if all the forecast errors were independent of each other. However, in practice this is unlikely to be the case. In the above situation there is very little experience of forecasting and the costs involved are sufficiently high to discourage much purely experimental work. A natural reaction is to simply ignore the error analysis side of forecasting and hope that the forecast values will be reasonably good. There is clearly no hope of giving confidence intervals for the forecast of $\hat{s}_t$. While this may all be true it is still nonetheless possible to investigate the forecasting method and say quite a lot about the behaviour of the forecasts. One line 1 of Table 15.5 we give the forecast values of all the variables. On line 2 the values of $\hat{n}_{t-1}$, $\hat{a}_{t-1}$ and $\hat{l}_{t-1}$ have all been set at what in the practicalities of the situation are thought to be the most extreme values in terms of making n_t large. These are perhaps our guesses at the 95 per cent confidence limits. The values of $\hat{p}_t$ and $\hat{r}_t$ are left alone. On line 3 we take the line 1 value of $\hat{n}_t$ but let $\hat{p}_t$ and $\hat{r}_t$ take on low values. The sizes of the schools involved suggest only small drops in $\hat{p}$ and $\hat{r}$ are likely and also puts clear upper values on them. On line 4 we take these low values with the high value of $\hat{n}_t$ of line 2. By this type of exercise we can begin to appreciate the effects on the final answer of variations in the forecasts. If one is willing to make experimental assumptions about the distributional behaviour of the forecasts, e.g. that $\hat{n}_{t-1}$, $\hat{a}_{t-1}$, $\hat{l}_{t-1}$ are independent with means as on line 1 and standard deviations 100, 50 and 10, respectively, then we can go even further. Clearly, $\hat{n}_t$ will have mean 2,350 and standard deviation

$$\sqrt{(100^2 + 50^2 + 10^2)} = \sqrt{(12,600)} \simeq 112$$

To find out something of the behaviour of $\hat{a}_t$ we must carry out a simulation. We could, for example, simulate $\hat{n}_t$ from a normal distribution $N(2,350, 12,600)$ and let $\hat{p}_t$ and $\hat{r}_t$ be normal $N(0.05, 0.01)$, with a correlation of, say, 0.5. On this basis many values of $\hat{s}_t$ could be generated and their mean, standard deviation and general distribution investigated. The essential part of this type of investigation is the need for the expert in the particular area to quantify his experience and intuition so that the sensitivity analysis can be realistic in its comparisons of the various possible futures.

Another aspect of a sensitivity analysis is the possibility of comparing the results of various possible decisions or assumptions. Suppose, for example, there is the possibility of a new housing estate in the area which would generate an additional 400 'arrivals'. The effect of this on the forecast mean situation of line 1 and the extreme situation of line 4 are shown in lines 5 and 6, respectively. We have taken for our illustration of this approach a situation which, though complex relative to the previous situations discussed in this book, is comparatively straightforward. We can guess the magnitude and signs of the effects of various changes possible. In more involved situations this is no longer the case and more elaborate sets of assumptions have to be investigated in the sensitivity analysis.

References

Davis, R., and Huitson, A. (1967). A sales forecasting comparison. *The Statistician*, 17, 269–237.

Godolphin, E. J., and Harrison, P. J. (1975). Equivalence theorems for polynomial-projecting predictors. *J. Roy. Statist. Soc. B*, 37, 205–215.

Granger, C. W. J. and Newbold, P. (1973). Some comments on the evaluation of economic forecasts. *Applied Economics*, 5, 35–47.

Groff, G. K. (1973). Empirical comparison of models for short range forecasting. *Management Science*, 20, 22–30.

Marchant, L. J., and Hockley, D. J. (1970). A comparison of two forecasting techniques. *The Statistician*, 20, 35–64.

McKenzie, E. (1974). A comparison of standard forecasting systems with the Box–Jenkins approach. *The Statistician*, 23, 107–116.

Newbold, P., and Granger, C. W. J. (1974). Experience with forecasting univariate time series and the combination of forecasts. *J. Roy. Statist. Soc. A*, 137, 131–146.

Theil, H. (1958). *Economic Forecasts and Policy*. North-Holland Publishing Co., Amsterdam.

Theil, H. (1965). *Applied economic forecasting*, North-Holland Publishing Co., Amsterdam.

Wagle, B., Rappoport, J. Q. G. H., and Downes, V. A. (1968). A program for short-term sales forecasting. *The Statistician*, 18, 141–145.

Williams, W. H., and Goodman, M. L. (1971). Constructing empirical confidence limits for economic forecasts. *JASA*, 66, 752–755.

Chapter 16

Forecast control

16.1 Quality control in forecasting

We have seen that the essence of statistical forecasting is the extrapolation into the future of some structure in the past and present situation. If our assumption that the given structure will continue into the future is correct, we will get good forecasts but not perfect ones, since in the real world the structure will always contain some random element which is unpredictable. The forecast errors in such a situation will have some statistical distribution which remains constant. We will sometimes get large errors in this situation, but only with the probability given by this distribution. A situation which has this type of stability is said to be in 'statistical equilibrium'. If we have a stable structure and any fixed forecasting formula, the forecasting errors will be in this type of statistical equilibrium. Figure 16.1 shows a set of forecast errors from such a situation. An initial set of data yields a standard deviation of 130, using a zero mean, and a sample mean close to zero. The histogram of the errors suggests that a normal distribution would provide a reasonable model for the distribution of the errors and further study supports this. As there are a large number of observations, the sample standard deviation can be taken as a good estimate of the population standard deviation and the mean may reasonably be taken as zero. From the table of the normal distribution it is seen that 95 per cent of the observed errors should lie within 1.96 (approximately 2) standard deviations of the mean. Lines corresponding to these values, i.e. ± 260, are drawn in Figure 16.1. Such lines are called control limits. As long as the situation remains in statistical equilibrium the errors will only fall outside these two sets of limits on an average of 5 per cent of occasions. If, however, a change occurs in the structure, perhaps suddenly, perhaps gradually, the statistical equilibrium will be lost or a new equilibrium developed. In either case the statistical structure of the errors will change. This change may be revealed in many ways, by a

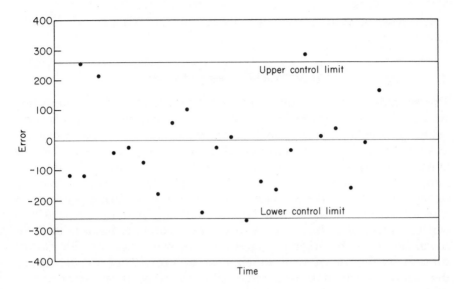

Figure 16.1   Control chart for error

change in the mean error or the error standard deviation and a consequent change in the proportion of observations outside the control limits. When such a change in the situation occurs the forecasting method needs to be modified. This modification may take the form of a change in the parameters of the forecasting formula, a complete change in the forecasting formula or just a restart of the method with new initial values.

The use of a graph as a simple visual check on statistical equilibrium, as in Figure 16.1, has much to recommend it. Unfortunately, in many forecasting applications the situation is rarely stable long enough for the calculation of control limits; nor, indeed, is this the most efficient way of carrying out a control of the forecast errors. For example, if there is a small change in the mean this will not show very clearly when only the individual observations are plotted on the chart of Figure 16.1. It is intuitively better in this case to combine the errors into some measure of the change in mean. This measure will have known statistical properties when the situation is in equilibrium. In general, such measures will be chosen so that for particular changes in structure there will be clear changes in the measure used. It is clear that for any given measure, such as the mean error over the available data, certain changes will show up clearly but others will have little effect. A sudden change in the mean will in time appear clearly as a change in the mean error. But a change in structure which takes the form of, say, some type of added periodic term may not have much effect on the mean error. In the next section we will consider some measures of error structure

248

which have proved particularly useful in the practice of forecast control.

16.2 Tracking signals for bias

The simplest way to keep a watch on the stability of a forecasting system is to use what has been called a 'tracking signal'. This is some quantity calculated from the forecast errors and such that when the situation is stable the tracking signal usually remains between specified limits, but when the situation ceases to be under control the tracking signal goes outside these control limits. The first suggested tracking signal was that of Brown (1959) and is based on using the sum of forecast errors. An alternative to this was suggested by Trigg (1964) which, though it is based on a very similar formula, has a number of advantages over the original suggestion of Brown. Trigg's tracking signal is based on the use of a smoothed mean of errors. Thus if e_t represents the forecast error at time t and $\tilde{e}_t$ the smoothed mean error at that time, then

$$\tilde{e}_t = (1 - b)e_t + b\tilde{e}_{t-1}$$

where b is the smoothing constant having a value fairly close to one. Assuming a stable situation in which e_t has expectation zero and variance σ^2 and taking the expectation and variance of both sides of this equation gives, for large t,

$$E(\tilde{e}_t) = 0$$

and

$$\mathrm{Var}(\tilde{e}_t) = \frac{1-b}{1+b}\,\sigma^2 = \sigma_{\tilde{e}}^2$$

It is assumed here that the e_t are approximately uncorrelated. This assumption usually holds in practice, but it should be checked when using such a tracking system.

The smoothed error is used to provide a quality control procedure for the forecast errors. If the errors can be treated as approximately normally distributed, then 95 per cent of the values of the smoothed error should lie within approximately $\doteq 2\sigma_{\tilde{e}}$. However, σ, and hence $\sigma_{\tilde{e}}$, is not known and has to be estimated. Rather than estimate σ^2 by the classical root mean square estimate, it is easier in this situation to use an estimate that involves the minimum of calculation and can easily be recalculated as each new error is obtained. Brown has suggested the use of a smoothed version of the mean deviation of errors, which is called the mean absolute deviation (MAD). Denoting the value of this at time t by MAD_t, then the formula for calculating the MAD is

$$\mathrm{MAD}_t = (1 - b)\,|e_t| + b\,\mathrm{MAD}_{t-1}$$

where $|e_t|$ denotes the absolute (signless) value of e_t. The smoothing constant is the same as that used for e_t, for reasons that will become apparent later. It can be shown that for a normal distribution

$$E(|e_t|) = \sqrt{\frac{2}{\pi}}\,\sigma$$

and Brown (1959) shows that for the normal, exponential, rectangular and triangular distribution

$$E\,(|e_t|) \doteqdot 0.8\,\sigma \doteqdot \sigma/1.2$$

Since $E\,(\mathrm{MAD}_t) = E(|e_t|)$ the quantity $1.2\,\mathrm{MAD}_t$ provides an estimate of σ_e. Thus $2\,\sigma_e$ control limits would become $\doteqdot 2.4\,\mathrm{MAD}_t$. These control limits will vary as MAD_t varies and will be recalculated everytime a new observation is obtained. Rather than have variable limits the smoothed error is divided by the quantity MAD_t, giving a quantity which will have fixed control limits. Thus the tracking signal used is

$$T_t = \tilde{e}_t/\mathrm{MAD}_t$$

To obtain the control limits for T_t we note that, using the 95 per cent control limits and the relations derived above, the situation is assumed to be in a state of equilibrium when

$$-2\,\sigma_{\tilde{e}} < \tilde{e}_t < 2\sigma_{\tilde{e}}$$

$$-2\,\sqrt{\frac{1-b}{1+b}}\,\sigma_e < \tilde{e}_t < 2\,\sqrt{\frac{1-b}{1+b}}\,\sigma_e$$

$$-2\,\sqrt{\frac{1-b}{1+b}}\,1.2\mathrm{MAD}_t < \tilde{e}_t < 2\,\sqrt{\frac{1-b}{1+b}}\,1.2\,\mathrm{MAD}_t$$

$$-2.4\,\sqrt{\frac{1-b}{1+b}} < T_t < 2.4\,\sqrt{\frac{1-b}{1+b}}$$

If T falls outside these limits then there is a suggestion that all is not well. We do not put it more strongly than this, since 5 per cent of observations will by chance fall outside these limits even when all is well. One assumption behind this calculation is that MAD_t gives a good approximation to σ_e. This is only reasonable when t is large and b is close to one.

A problem associated with the use of this tracking signal is that both e_0 and MAD_0, the initial values in the recurrence relations, need specifying before the data e_1, e_2 become available. A reasonable value for e_0 would be zero and for MAD_0 would be $\sqrt{2/\pi}\,\hat{\sigma}_0$, where σ_0

would be an estimate of the error variance. Some guide as to the value to put for $\hat{\sigma}_0$ can be obtained from knowledge of the forecasting formula and of the probable data. for example, if the recurrence form of exponential smoothing is to be used on the data, i.e.

$$\tilde{x}_t = (1 - \gamma)x_t + \gamma\tilde{x}_{t-1}$$

then the variance of $\tilde{x}_t$ is

$$\sigma_{\tilde{x}}^2 = \frac{1 - \gamma}{1 + \gamma} \sigma_x^2$$

assuming independent observations with locally constant mean and variance σ^2. Hence the error variance is given by

$$\mathrm{Var}(e_t) = \mathrm{Var}(x_t - \tilde{x}_{t-1})$$

$$= \sigma_x^2 + \frac{1 - \gamma}{1 + \gamma} \sigma_x^2$$

$$= \frac{2}{1 + \gamma} \sigma_x^2$$

By carrying out this analysis we have not gained any information but have simply replaced the requirement for a guess at MAD_0 by a guess at σ_x^2, which is usually more within one's intuitive grasp of the situation. If by comparing the variable being forecast with other similar variables we estimate σ_x by $\hat{\sigma}_x$, then the initial value for MAD is

$$\mathrm{MAD}_0 = \frac{2\sigma_x}{\sqrt{\{\pi(1 + \gamma)\}}}$$

A better solution to this problem is to defer the commencement of the scheme until enough actual data are available to provide an estimate of the mean deviation of errors and to use this value, i.e.

$$\mathrm{MAD}_0 = \sum_{\substack{\text{available} \\ \text{data}}} |e|/(\text{number of observations})$$

We can, however, avoid these starting value problems by using a slightly different form of tracking signal.

The exponential smoothing recurrence relation which has the form

$$\tilde{z}_t = (1 - b)z_t + b\tilde{z}_{t-1}$$

for some sequence z_t, is basically an approximation, when t is large, to the discounted average

$$\tilde{z}_t = \sum_{r=0}^{t-1} b^r z_{t-r} \bigg/ \sum_{r=0}^{t-1} b^r$$

Using this exact form in the tracking signal gives

$$T'_t = \frac{\displaystyle\sum_{r=0}^{t-1} b^r e_{t-r} \bigg/ \sum_{r=0}^{t-1} b^r}{\displaystyle\sum_{r=0}^{t-1} b^r \mid e_{t-r} \mid \bigg/ \sum_{r=0}^{t-1} b^r}$$

$$= \sum_{r=0}^{t-1} b^r e_{t-r} \bigg/ \sum_{r=0}^{t-1} b^r \mid e_{t-r} \mid$$

$$= S_t(e)/S_t(\mid e \mid)$$

where $S_t(\ )$ represents the exponentially weighted sum, e.g.

$$= S_t(z) = \sum_{r=0}^{t-1} b^r z_{t-r}$$

The advantage of this form for the tracking signal is that there is no requirement for starting values in recurrence relations; T'_t depends only on the data available and not on our estimates. Strangely enough, though this is the 'exact' form it is easier to calculate since quantities like $S_t(z)$ have a very simple recurrence form, namely

$$S_t(z) = z_t + b S_{t-1}(z)$$

The starting value for this recurrence relation is simply $S_0(z) = 0$ so that $S_1(z) = z_1$. Obviously if t is large then T'_t and T_t are equivalent.

Let us examine the properties of T_t and T'_t in a little more detail. Suppose that the process was completely out of control from the start so that all the errors had the same sign. If the sign was positive there would be no difference between $e_1, e_2, \ldots, e_t$ and $\mid e_1 \mid, \mid e_2 \mid, \ldots, \mid e_t \mid$, and hence between $\tilde{e}_t$ and MAD_t, so $T_t = 1$. If the sign was negative then $e_i = -\mid e_i \mid$, $i = 1, \ldots, t$, and so $T_t = -1$. As these are the two extreme possibilities it is clear that T, and equivalently T', must lie in the interval $(-1, 1)$. This useful condition is the result of using b as the smoothing constant for both $\tilde{e}_t$ and MAD_t.

In the derivation of the control limits it was pointed out that e_t was assumed to be normal and MAD_t was treated as a constant; this implies the assumption that T is normal, which cannot be the case with T lying between limits. The validity of a normal approximation for T must be checked and it is probably easiest to do this by simulation. Thus we may generate a long sequence of random variables e having the sort of property we expect from our actual situation when it is stable. The sequence of T values calculated from the e values is then examined and a cumulative frequency polygon constructed. From this we can read off values of T which will only be exceeded with any required probability. A discussion of the distribution of T is given in Brown (1963).

16.3 Cumulative sum methods

It is difficult to detect small changes in the mean error using the type of control chart discussed in section 16.1. One way of improving the sensitivity is to base the control chart on the averages of several errors instead of on the individual errors. If we could only guess where a change in mean had occurred, then the average of all the errors obtained since the change would be most sensitive in indicating that the change had occurred. Unfortunately, we do not know when the change occurred, if it did, so we do not know how many observations to include in our average. One way of avoiding this problem is to have a whole battery of control procedures based on the averages of all past sets of errors, one using just the last error, one using the average of the last two, one the average of the last three, and so on. Each of these averages will be the most sensitive available for detecting a sudden change in the underlying mean error, provided the change in mean occurred immediately before the earliest error used in that particular average. If we have say 100 observations, this process effectively involves the construction of 100 control procedures. This is obviously impractical but the basic idea does lead to some very useful methods of control. The first modification required to obtain a practical method is to replace the average errors by the sums of errors. These are easier to calculate and are easily related to each other. Thus the quantities used as the basis for the procedure are

$$S_1 = e_t$$
$$S_2 = e_t + e_{t-1}$$
$$S_3 = e_t + e_{t-1} + e_{t-2}, \text{etc.}$$

which are called cumulative backward sums.

A further modification to the procedure becomes apparent on noting that, for example, the quantity S_{20} would provide the most efficient measure that we could use to detect a step-change in the underlying mean which occurred 20 time units ago. However, for our purposes we would hope that any change of significant magnitude which occurred would certainly be detected long before 20 time units after its occurrence. In practice, then, we will only be interested in the first few values of S. In a discussion of this method in the ICI monograph on short-term forecasting (Coutie and others, 1966), the use of S_1, $S_2, \ldots, S_6$ is suggested. The control procedure is thus to specify control limits $\pm L_1, \pm L_2, \ldots, \pm L_6$ for these sums and take the process as being in equilibrium as long as all six S_i lie within their appropriate limits. It is possible to calculate values for $L_1, \ldots, L_6$ so that each S has some specified probability of falling outside the limits when the process is under control. The six controls are actually used as a group so that these individual probabilities are not very informative, particularly since the values of $S_1, \ldots, S_6$ are not independent of each other. Thus the problem of choosing theoretically a set of values for the L values is extremely complicated. The procedure usually adopted is to express the control limits L_i ($i = 1, \ldots, 6$) as functions of i which involve some simple parameters and then to choose the best parameters by a simulation study. Obviously the values of L will increase with the number of errors summed, i.e.

$$L_1 < L_2 < \ldots < L_6$$

A very simple form of control limit is obtained by taking L_i as a linear function

$$L_i = W\sigma(i + h)$$

where σ is the standard deviation of the errors and W and h are the parameters to be chosen. To choose these parameters, simulation methods can be used. Thus we generate two sets of data from a normal distribution, or whatever distribution is appropriate, with the given σ and means zero and d, where d is a change in the mean of a magnitude that should be detected fairly rapidly, e.g. $d = 1.5\,\sigma$. These data are then used in the control scheme for a range of values of w and h. The measure of control of clearest meaning here is the average run length (ARL), which is the average number of observations after a change in mean before the control system indicates a possible lack of control. If a real set of data giving forecasts with average error close to zero is available, these data and the data with d added to each observation can be used in place of the simulated sets of data. Results from such a simulation are given in the ICI monograph referred to above. By way of example, suppose $W = 0.4$ and $h = 3.4$. Then when the situation is under control the average run length is 9, so that with monthly data we will unnecessarily review the situation once in 9 months. When, however, a change in mean of magnitude $1.5\,\sigma$ occurs, a lack of control will be indicated after about 2 observations.

The above describes the principle of the method, but for practical operation a method can be devised by which only four quantities need be recorded to use in the scheme. This suprising point is made in the paper by Harrison and Davies (1964). The essence of the scheme is that in looking for points which might cross the control limits one need only look at the sum which is closest to the limits. A recurrence relation is devised which enables this closest point to be identified at each stage, without having to investigate points which are nowhere near the limits.

16.4 A Tracking Signal for autocorrelation

The tracking signals of the previous section were designed to indicate the occurrence of any significantly large local bias. It is important, particularly with stochastic models, to check the forecast errors for the occurrence of autocorrelation between successive errors. Ideally the errors will be independent. There is therefore need for a tracking signal to indicate any large local autocorrelation as well as bias. The autocorrelation of lag 1 is the natural one to study, but others might be checked as well. A simple tracking signal for autocorrelation can be based on the Von-Neuman ratio. This is normally used as a test for lack of independence. It is based on the mean square successive difference, i.e. on

$$\delta^2 = \frac{1}{n-1} \sum_{i=1}^{n-1} (e_{i+1} - e_i)^2$$

If the successive errors are independent, this has an expected value of $2\sigma_e^2$. The error variance σ_e^2 can be estimated by

$$S^2 = \frac{1}{n} \sum_{i=1}^{n} (e_i - \bar{e})^2$$

Thus the ratio $K = \delta^2 / S^2$ is approximately 2 if the e values are independent. If there is a large positive autocorrelation, the e_{i+1} values will tend to be close to the e_i values so δ^2 will be smaller. This can be turned into a tracking signal by replacing all the above averages by the corresponding discounted quantities. Thus

$$\delta_t^2 = \sum_{r=0}^{t-2} b^r (e_{t-r} - e_{t-1-r})^2 \left/ \sum_{r=0}^{t-2} b^r \right.$$

$$S_t^2 = \sum_{r=0}^{t-1} b^r (e_{t-r} - \tilde{e}_{t-r})^2 \left/ \sum_{r=0}^{t-1} b^r \right.$$

where $\tilde{e}_{t-r}$ is the exponentially weighted average of the e values at time $t - r$. Thus, for large t, sufficient to let

$$\sum_{0}^{t-1} b^r \simeq \sum_{0}^{t-2} b^r$$

we can define the tracking signal for autocorrelation by

$$K_t = \frac{\displaystyle\sum_{r=0}^{t-2} b^r (e_{t-r} - e_{t-1-r})^2}{\displaystyle\sum_{r=0}^{t-1} b^r (e_{t-r} - \tilde{e}_{t-r})^2} = \frac{\delta_t^2}{S_t^2}$$

As usual, numerator and denominator can be evaluated using recurrence relations. Control limits for K_t are best found using simulations of both independent errors and of errors showing autocorrelations of the various magnitudes that one would wish to detect in practice.

References

Brown, R. G. (1959). *Statistical Forecasting for Inventory Control*. McGraw-Hill, New York.

Brown, R. G. (1963). *Smoothing, Forecasting and Prediction of Discrete Time Series*. Prentice-Hall. Englewood Cliffs, New Jersey.

Coutie, G. A., Davis, O. L., Hassall C. H., Miller, D. W. G. P. and Morrell, A. J. H. (1966). *Short Term Forecasting*. ICI Monograph No. 2. Oliver and Boyd. Edinburgh.

Harrison, P. J. and Davis, O. L. (1964). The use of cusum techniques for the control of routine forecasts of product demand. *Op. Res. Quarterly*, **12**, 325–333.

Trigg, D. W. (1964). Monitoring a forecasting system. *Op. Res. Quarterly*, **15**, 271–274.

Chapter 17

Two-stage forecasting

17.1 Introduction

There are three basic aspects of applied forecasting: first, that of analysis of past data and the construction of models for this; second, the analysis of the use to which the forecast is to be put and again the appropriate model building; and third, the bringing together of the two to obtain the best forecasting method for the given application. Having said this we must acknowledge that this is the ideal. At the opposite extreme to the ideal, data are poured into one end of a computer manufacturer's forecasting program and whatever comes out at the other end is used as the forecast. There are several factors that tend to encourage people to take an approach that is less than ideal.

(a) Manufacturer's programmes are frequently tied to specific forecasting methods and give the user very little opportunity to analyse his data.
(b) The amount of work involved in the ideal approach is often prohibitive. For example, it often occurs that a forecast of some quantity is needed for several different purposes in a firm. The ideal would be to use separate forecasts for each purpose since the criteria will be different for different purposes, but the effort required discourages this and a single forecast is probably used for all purposes.
(c) There are only a few well-tried and understood methods of forecasting, so there is a natural tendency to keep to these rather than to launch out with tailor-made methods for particular types of data and applications.

An implication of our failure to take the theoretically ideal approach is that our forecasts are not the best that we could obtain. The question thus arises as to whether we can improve our forecasts, without all the work involved in the ideal approach to forecasting. In this chapter we

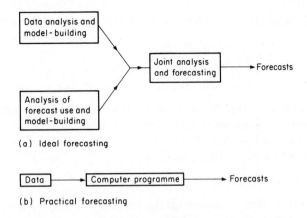

(a) Ideal forecasting

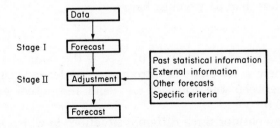

Figure 17.1   The theory and practice of forecasting

will introduce the idea of two-stage forecasting. This seeks to improve forecasts by means of a compromise between the more usual approach and the ideal. The forecast obtained by some standard method is regarded as the first stage in the forecasting process. We will assume that the properties of forecasts obtained from this method, under stable conditions, are known to us. The second stage consists of an adjustment of this stage I forecast to obtain a new, stage II, forecast which is more appropriate to the application or data on hand.

The basis from which the second-stage forecast can be constructed include:

(a) the use of statistical properties of past forecast errors;
(b) the use of external information;
(c) the use of other forecasts;
(d) the use of different forecasting criteria.

By the use of any of these approaches we may seek to produce a final forecast that is more appropriate to the situation but yet still uses the standard methods to obtain the basic forecast. In fact we may

Figure 17.2   Two-stage forecasting

produce several forecasts for different purposes from the same stage I forecast.

17.2 The use of forecast errors

A natural way to seek to improve forecasts is to see whether the forecast errors still contain any structure that can be used. Let us denote the stage I forecast of x_{t+h} by $\hat{x}_{t+h}$, the error by e_{t+h} and the stage II forecast by $\hat{y}_{t+h}$. By definition,

$$\hat{x}_{t+h} = x_{t+h} + e_{t+h}$$

If now by using the structure in the forecast errors we may obtain a forecast, $\hat{e}_{t+h}$, of e_{t+h}, we can define a second-stage forecast by

$$\hat{y}_{t+h} = \hat{x}_{t+h} + \hat{e}_{t+h}$$

The forecast error, z_{t+h}, of the second stage is thus

$$z_{t+h} = x_{t+h} - \hat{y}_{t+h}$$

$$= e_{t+h} - \hat{e}_{t+h}$$

by substituting from the above equation, It is clear that we will obtain reduced forecast errors provided the forecast $\hat{e}_{t+h}$ is a reasonable forecast of e_{t+h}. There are usually two features that we would like to improve by the use of $\hat{e}_{t+h}$. Firstly we would like to reduce any bias that there may be in the stage I forecasts. We can do this completely if the forecast $\hat{e}_{t+h}$ is such that

$$E(\hat{e}_{t+h}) = E(e_{t+h})$$

for then $E(z_{t+h}) = 0$. Secondly, we will want to reduce the forecast mean square error, or variance if the forecasts are unbiased. Consider the latter case. Denote the variances of z_{t+h}, e_{t+h} and $\hat{e}_{t+h}$ by σ_z^2, σ^2 and $k^2 \sigma^2$, respectively, and the correlation between $\hat{e}_{t+h}$ and e_{t+h} by ρ. Then

$$\sigma_z^2 = \sigma^2 + k^2 \sigma^2 - 2\sigma k\sigma \rho$$

For σ_z^2 to be less than σ^2 we must have

$$\rho > k/2$$

Thus, for example, if the variance of $\hat{e}_{t+h}$ is the same as that of e_{t+h}, $k = 1$, then there is no point in introducing the second stage unless the correlation between the forecast $\hat{e}_{t+h}$ and the actual error e_{t+h} is greater than 0.5.

Let us now consider three different situations in which we can make adjustments on the basis of evidence from previous forecasts.

17.2.1 Adjustment for bias

The methods of this section are appropriate where an examination of the statistical properties of the forecast errors show some systematic bias that can be used to adjust the forecast. Having observed such a feature, a first step would be to examine the stage I forecasting method to see if the adjustment might be made there without the use of stage II. However, it is sometimes best to let well alone and correct the forecast with a second stage. Such a situation might occur where the stage I forecast is based on judgement. For example, forecasts of sales are sometimes based on the judgement of a forecasting committee. For reasons of strategy, possibly unconscious, the committee may over-estimate sales. Rather than try to interfere with the functioning of this committee, a better forecast could be obtained by reducing their forecast by an adjustment based on an examination of bias in past forecasts. One would thus have

Adjusted forecast = stage I forecast + estimated bias

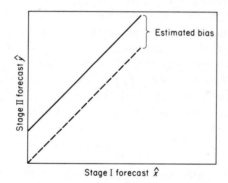

Figure 17.3   Adjustment for bias

In graphical form the method is shown in Figure 17.3.

As an illustration of this technique, we consider two practical examples. These examples also illustrate elaborations of the basic idea of two-stage forecasting that may be used in other applications of two-stage forecasting.

Example 1

The method discussed here was developed to provide forecasts of the number of telephone calls arriving at a switchboard during different half-hourly periods during a working week. The total time required to answer these calls was also forecast. The data showed very large systematic variations during the day and also between the days of the week. There were also odd days where, for unforeseeable reasons, the load was very heavy. The usual way to forecast such a situation might be to fit some trigonometric model to the data, perhaps with the use of some discounting procedure. Alternatively, some modified form of the classical methods for seasonal adjustment could have been used. However, both these methods would have required considerable computation. The system required had to be used by someone without any mathematical ability, and it had to give an answer by hand with something like a minute's effort. It had also to be very sensitive to pick out the odd day with extra high or low loads. In addition it was necessary for the system to follow changing patterns in the distribution of load within the week. Such changes occurred sometimes during different seasons of the year. Within this context any attempt to find the best models or forecasting formulae with smallest mean square error would have inevitably led to methods that were too complicated to work. Instead, attention was focused on modifying the simplest methods available.

The method devised was based on the technique of exponential smoothing in its simplest form. As this technique was not appropriate to data with periodic form, it was applied in two stages. An advantage of the method was that the preliminary stage I forecast could be made a week ahead and the final adjustment made on the actual day an hour in advance of the half-hour being forecast. The method used in stage I was to forecast separately for each of the 85 half-hourly periods in the working week, using for each period only data from that period in previous weeks. Thus exponential smoothing was applied to 85 separate series. The nomogram given in Chapter 4 was used to simplify the practice.

This method gave adequate forecasts most of the time. There were, however, occasional days when the load was particularly high. This produced a systematic positive bias in the forecast errors on such a day. This bias was forecast one hour in advance by exponentially smoothing the forecast errors from stage I as they arose during the day. Thus if $\tilde{x}_{10}$ was the stage I forecast, made the previous week, of the tenth half-hour in the day and $\tilde{e}_8$ was the exponentially smoothed error obtained at the eighth half-hour, then at that time a revised forecast would be

$$\tilde{y}_{10} = \tilde{x}_{10} + \tilde{e}_8$$

Such an adjustment was very effective on those days when something special happened. It was found also that because of trends during the year there was usually a small bias present on most days. If the stage I forecast was always adjusted by a multiple, in this case 0.5, of the smoothed error, there was found to be an overall 5 to 10 per cent improvement in the forecast root mean square error.

Example 2

A large firm produces at the end of each year a planned sales for each week of the following year. This plan was based on fitting a linear regression model to log sales data and then making various adjustments for holidays, etc. A study of several years' data indicated that this was a satisfactory model. The use of log sales in the model suggested that errors be defined as percentage deviations from plan. Thus if p_t is the planned sales for week t, x_t the actual sales and d_t the percentage deviation, then by definition

$$x_t = p_t(1 + d_t/100)$$

A study of past values of d_t indicated that in any year there were runs of deviations above and below zero and that there was an overall bias away from zero. If we manage to forecast the future value, d_{t+h}, by $\hat{d}_{t,h}$, then it is reasonable that a revised plan will be given by

$$\hat{x}_{t,h} = p_{t+h} (1 + \hat{d}_{t,h}/100)$$

p_{t+h} being known. As the plan is used to decide on capital investment and manpower establishments at peak periods, it is clearly desirable to seek to produce such revisions. A practical problem here is that the lead time h, over which these decisions have to be made, may be so large that our forecast $d_{t,h}$ is poor and the revised plan would have worse properties than the original. This would require careful investigation in terms of costs of errors in the plan. Assuming that $\hat{d}_{t,h}$ is sufficiently good forecast, for our planning the deviation, d^*_{t+h}, of x_{t+h} from the revised plan is given by

$$d^*_{t+h} = (d_{t+h} - \hat{d}_{t,h})/(1 + \hat{d}_{t,h}/100)$$

Table 17.1 gives some properties of the quantities d, $\hat{d}$ and d^* from the situation described. As the values d are obtained from deviations from the yearly plans, there are no systematic changes with h. The forecasts $\hat{d}$ naturally get worse as h increases, however they have reduced the bias in d to such a large extent that the mean square of d^* is still appreciably reduced over a nine-week lead time.

Table 17.1 Properties of 'deviation from plan' based on 19 weeks' data

		Lead time in weeks							
		1			5			9	
	Mean	Standard deviation	Mean square	Mean	Standard deviation	Mean square	Mean	Standard deviation	Mean square
d	7.18	10.5	161.3	6.81	10.4	154.5	6.85	10.5	156.3
$\hat{d}$	0.07	10.7	114.5	−0.39	11.2	123.6	−0.50	11.3	128.0
d^*	0.19	10.0	100.3	−0.19	10.5	108.9	−0.28	10.6	112.0

It is not appropriate to discuss the forecasting methods used in this particular example, except to point out that in this type of situation we are often particularly interested in runs of deviations either above or below zero and thus it may be advisable only to introduce a revised, second-stage plan if there is some indication that such a run is taking place. Or, alternatively, we may have a constantly used second-stage revision of plan to allow for overall bias and, when appropriate, introduce a third stage to allow for runs.

17.2.2 The use of autocorrelation in errors

In section 15.2 (viii) and Figure 15.9 we looked at plots to show relationships between successive errors. If such relationships are found, they can be used to forecast future errors and thus provide a second-stage forecast. The most common relationship is an auto-correlation between successive errors e. The simplest model for this is the autoregressive model of section 7.1, i.e.

$$e_t = \phi e_{t-1} + \epsilon_t$$

The parameter ϕ is the autocorrelation of lag 1. This can be estimated

Table 17.2 Two-stage forecasting — autoregression adjustment

Autocorrelation of lag 1 = 0.48 = $\hat{\phi}$
Second-stage forecast $\hat{y}_t = \hat{x}_t + \hat{\phi} e_{t-1}$

1 x_t	2 $\hat{x}_t$	3 (1) − (2) e_t	4 $\hat{e} = \hat{\phi} e_{t-1}$	5 (2) + (4) $\hat{y}$	6 (1) − (5) $e_{\hat{y}}$
		—			—
336.3	352.0	−15.7	−3.2	348.8	−12.5
339.5	363.6	−24.1	−7.5	355.8	−16.3
315.3	377.8	−62.5	−11.6	366.2	−50.9
359.0	384.7	−25.7	−30.0	354.7	4.3
364.3	382.3	−18.0	12.3	370.0	−5.7
353.0	347.6	5.4	−8.6	339.0	14.0
366.0	365.0	1.0	2.6	367.6	−1.6
369.0	372.4	−3.4	0.5	372.9	−3.9
363.0	367.7	−4.7	−1.6	366.1	−3.1
350.8	368.4	−7.6	−2.3	366.1	−15.3

Comparison					
	Mean absolute error	17.82			12.76

Example here uses $\hat{x}$ obtained by a regression method.

from past information or we can continually update our estimate as illustrated in Table 7.5. Thus, given an estimate $\hat{\phi}$, we have for the stage II forecast (see Table 17.2)

$$\hat{y}_{t+1} = \hat{x}_{t+1} + \hat{\phi}e_t$$

17.2.3 Regression adjustment methods

Suppose that when we plotted a set of forecasts and observations on the forecasts—observation diagram we obtained not the line of perfect forecasts but some other perfect curve $x = f(\hat{x})$, as in Figure 17.4(a). If we now define a second-stage forecast by

$$\hat{y}_{t+h} = f(\hat{x}_{t+h})$$

then $\hat{y}_{t+h}$ will equal x_{t+h} and we will obtain perfect forecasts. Obviously, in practice, we will not have anything remotely like a perfect curve. It does, however, sometimes occur that the forecast—observation diagram reveals a pattern of points through which we can fit a suitable line by standard statistical methods. The most common

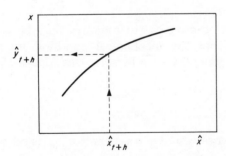

(a) A perfect relation on the forecast-observation diagram

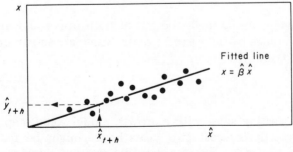

(b) A case of overestimation

Figure 17.4   Adjustment using regression

pattern corresponds to a bias whose magnitude increases with the magnitude of the mean. Thus we have a regression line through the origin but not at 45°, as is illustrated in Figure 17.4(b). The slope of this line is estimated, using least squares, by

$$\hat{\beta} = \Sigma x\hat{x} / \Sigma \hat{x}^2$$

Thus the second-stage forecast is

$$\hat{y} = \hat{\beta}\,\hat{x}$$

A study of adjustments, based on linear regression, applied to economic forecasting is given in Theil (1958) and (1965).

17.3 The use of external forecasting information

We consider in this section the problem of adjusting forecasts to allow for external information that can be expressed as simple mathematical or statistical restrictions on the forecast. These methods are, for the most part, based on the ideas in a paper by Zellner (1963), though the approach is somewhat different. Consider a few examples.

Example 1

The stage I forecast of the output of a quarry may be greater than the maximum amount that the mechanical handling equipment can cope with, x_{max} say. In general we could write (see Figure 17.5a)

$$\text{Stage II forecast} = \begin{cases} \hat{x} & \text{if } \hat{x} < x_{max} \\ x_{max} & \text{if } \hat{x} > x_{max} \end{cases}$$

A similar situation may arise where a lower limit x_{min} is set so that

$$\text{Stage II forecast} = \begin{cases} \hat{x} & \text{if } \hat{x} > x_{min} \\ x_{min} & \text{if } x < x_{min} \end{cases}$$

or, alternatively, limits might be put at both upper and lower ends of the expected range of the forecast value. These are shown graphically in Figure 17.5(a) and (b).

Example 2

When forecasting sales with a growth model some scepticism about the rapid growth forecast two years hence might be felt. If 'expert opinion' thought x_{max} was a value that was unlikely to be exceeded, then a compromise between $\hat{x}$ and x_{max} might be made when $\hat{x}$ exceeded x_{max}. In mathematical terms we might put

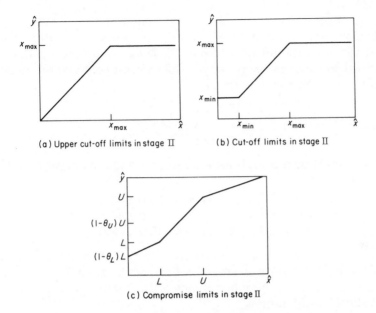

(a) Upper cut-off limits in stage II   (b) Cut-off limits in stage II

(c) Compromise limits in stage II

Figure 17.5   Adjustment using external information

$$\text{Stage II forecast} = \begin{cases} \hat{x} & \text{if } \hat{x} < x_{\max} \\ \theta\hat{x} + (1-\theta)x_{\max} & \text{if } \hat{x} > x_{\max} \end{cases}$$

The parameter θ $(0 < \theta < 1)$ would be made small if strong reliance was put on the expert opinion rather than the forecasting method. If the converse, θ would be put close to 1.

Example 3

In some cases the external information can be put in the form of a distribution for the future observation x_t. That is to say that, apart from knowledge of the stage I forecast, we have a detailed belief in the possible values of x_t in the form of a distribution function $p(x_t)$. The expectation of this distribution, which we might call the *a priori* distribution of x_t, provides a single forecast of x_t. Here, however, we wish to use knowledge of the distribution to modify the stage I forecast. For any distribution we may find a pair of values $(L.U)$ such that there is a small probability α of observations lying outside the interval $(L.U)$. We may choose L and U so that the probabilities are $\alpha/2$ of observations lying below L and $\alpha/2$ of observations lying above U. For example, if $p(x)$ is a normal distribution with mean μ and standard deviation σ, there are probabilities of 0.025 of observations lying below $L = \mu - 1.96\,\sigma$ and 0.025 of observations lying above $U = \mu + 1.96\,\sigma$.

If the stage I forecasts lie in the outer tails of our *a priori* distribution, as indicated by L and U, it seems reasonable that the stage II forecasts should seek to pull the forecasts in towards the mean. This could be done in many ways. The forecast could be set equal to L or U if it exceeded the bounds, i.e.

$$\text{Stage II forecast} = \begin{cases} L \text{ if } \hat{x} < L \\ \hat{x} \text{ if } L < \hat{x} < U \\ U \text{ if } \hat{x} > U \end{cases}$$

or it may simply be a compromise between x and the bound L or U. Thus

$$\text{Stage II forecast} = \begin{cases} \theta_L \hat{x} + (1 - \theta_L)L \text{ if } \hat{x} < L \\ \hat{x} \qquad\qquad \text{ if } L < \hat{x} < U \\ \theta_U \hat{x} + (1 - \theta_U)U \text{ if } \hat{x} > U \end{cases}$$

where θ_L and θ_U are chosen constants. For $\theta_L = \theta_U = 0$ this reduces to our first rule. The forecasts are, of course, of the same form as those in Examples 1 and 2, but here our choice of the limits is based specifically on statistical considerations.

If the reasoning on which $p(x)$ is based has more evidence on, say, the upper range of values, we might choose U so that the probability of exceeding U was very small and set θ_U also at a small value. This would have the effect of making it very unlikely that the stage II forecast will exceed U, which would reflect our confidence in this tail of the *a priori* distribution.

Studies of the effects of cut-offs indicate that, provided the information leading to the choice of L, U(and θ) is reasonable, the stage II forecasts are significantly better than the stage I forecasts. The information has to be positively misleading before the two-stage forecasting ceases to be of value. This would occur, for example, if U were some way below the true expected value of the future observation.

17.4 Combining forecasts

There are often several reasonable methods of forecasting available for a particular set of data. Rather than choose the best, which is often only marginally the best, it is often worthwhile to seek to combine the methods. If the methods use different features of the underlying situation, then such a combination may produce better forecasts than either forecast separately. A combination may be best achieved by devising a new model that combines the relevant features underlying the forecasting methods considered. However, it will usually be easier simply to combine the forecasts themselves. Suppose that we have two unbiased forecasts, $\hat{x}_1$ and $\hat{x}_2$. A simple linear combination of these is the most natural way of combining them. Thus the new forecast is

$$\hat{y} = k\hat{x}_1 + (1 - k)\hat{x}_2$$

The new forecast $\hat{y}$ will be unbiased and will have a minimum mean square error for suitably chosen k. If we consider the forecast errors of the three forecasts, it is clear that

$$e_y = ke_1 + (1 - k)\, e_2$$

using an obvious notation. If the forecast errors of $\hat{x}_1$ and $\hat{x}_2$ are independent of each other, then $E(e_y^2)$ is minimum when

$$k = \frac{E(e_2^2)}{E(e_1^2) + E(e_2^2)}$$

(See Bates and Granger, 1959). In practice we would replace this by

$$k = \frac{\tilde{e}_2^2}{\tilde{e}_1^2 + \tilde{e}_2^2}$$

where $\tilde{e}^2$ is a local estimate of the expected error squared. Newbold and Granger (1974) found the average from the previous twelve forecasts gave satisfactory results over a range of data. The exponentially smoothed error squared provides another alternative for $\tilde{e}^2$.

A somewhat different approach (Bunn, 1975), leads ultimately to a form of scoring system for the choice of k. Thus if in the last n forecasts $\hat{x}_1$ has been the closest to the observed values on r occasions, then

$$k = \frac{a_1 + r}{a_1 + a_2 + n}$$

The parameters a_1 and a_2 are chosen on the basis of one's prior assessment of the merits of the two forecasts. Thus if they were thought to be equally good we would put $a_1 = 1$, $a_2 = 1$ so that initially $k = \frac{1}{2}$.

Table 17.3 gives an illustrative example of these two approaches. Note that in both methods k is an adaptive parameter that changes in time to give the most weight to the most successful of the two forecasting methods being combined.

As an example of a slightly different variation on this theme, let us consider a case where the reliability of forecasts varied with time. In forecasting future telephone traffic, staff from several hundred 'telephone regions' were asked to forecast the percentage growth in telephone traffic for a number of years into the future. The central authority made an independent forecast for the country arriving at a figure of a constant growth of, say, p per cent. over the years that were forecast. It was felt that over short lead times of a year or two the

Table 17.3 Section of table showing combination of forecasts: combined forecast 1 uses the first method described, combined forecast 2 the second

Observation	Forecast 1	Forecast 2	Combined Forecast 1	Combined Forecast 2
97	115	120	116.9	ʼ115.7
102	106	94	101.5	104.7
96	79	74	ʼ77.2	78.6
100	75	108	87.2	77.0
84	68	87	78.2	73.3
MSE observations	14.6	14.3	12.7	13.9

locally made forecasts would be moderately good, but for longer lead times the p per cent. would be the best figure to use. It was also assumed *a priori* that growth rates for all regions should be positive. Using these assumptions the central authority adjusted all the regional forecasts. Denoting the regional forecasts of the percentage annual growth for a lead time of h years by $\hat{x}_h$ and the central authorities combined forecast by $\hat{y}_h$, the formula for the adjustment was

$$\hat{y}_h = \begin{cases} p & \text{if } \hat{x}_h < 0 \\ a^h \hat{x}_h + (1-a^h)p & \text{if } \hat{x}_h > 0 \end{cases}$$

where a is some constant in the range $(0,1)$. It is seen that this formula is similar to those above with k replaced by a^h. The effect of this form for k is that for large lead times y_h tends towards the value p. This adjustment procedure, with values of a based on trials with past forecasts, produced very large reductions in the average of the mean square errors over all the regions.

17.5 The use of cost and other criteria*

Most forecasting exercises are based on a minimum mean square error criteria. We have already referred to the situation where a single forecast based on this, or some other specific criteria, is used by different people or departments for different purposes. It may well be that each of these has a different criteria for what constitutes a good forecast. It may be that the costs depend on something different from the error squared or that the criteria may be framed in terms of minimizing probabilities rather than costs. In any such case, consideration should be given to the possibility of adjusting the initial forecast to produce a stage II forecast with better properties in terms of the specific criterion. The aim of this section is to give some illustrative examples of how this can be done. In all these examples we assume that we know the distribution of errors in the original forecast. Such information is usually available if the forecasting system has been in use for some time or if a reasonable amount of data is available for the investigation.

Example 1

Let the stage II forecast be of the form

$$\hat{y}_t = \hat{x}_t + \alpha$$

where α is some constant to be chosen to minimize the expected cost of the error in y_t. Assume that past records of $\hat{x}_t$ and x_t indicate that the forecast errors at stage I are normally distributed with zero mean and variance σ^2. The assumptions of normality and zero mean are brought in here to simplify the presentation; they are not necessary assumptions. The forecast error e of the stage II forecast will normally be distributed with mean $-\alpha$ and the same variance.

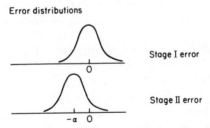

Error distributions

Stage I error

Stage II error

Figure 17.6   The effect of an additive adjustment

Many possible forms of cost function can be considered. It is clear that if the cost function and the error distribution are symmetric, with minimum cost when the error is zero, there will be no need to adjust the original forecast. There will only be a saving if the cost function is non-symmetrical. In such a case a biasing of the forecast, by putting $\alpha \neq 0$, will move the range of errors with highest probability into the region of the cost function involving lowest costs. It can be shown that for data based on a normal distribution and appropriate linear forecasting formulae the optimum predictor is obtained by such a procedure (see Granger, 1969).

Cost functions formed of linear sections are the simplest non-symmetric functions to manage mathematically. For example, the cost functions in Figure 17.7(a) and (b) are broken into four sections

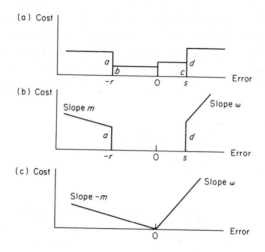

Figure 17.7   Some illustrative cost functions

representing minor under-or overprediction and major under- and overprediction. This type of cost function can be fitted to actual cost data as a reasonable approximation without requiring a great deal of information.

It is not proposed to go into mathematical detail in this section. Let us simply examine the general idea of the method. Let us denote the quantity $(e + \alpha)/\sigma$ by u; this will have a standard normal distribution, $N(0,1)$, with cumulative distribution denoted by $\Phi(u)$. That is to say, $\Phi(u_0)$ gives the probability that u is less than or equal to u_0. If we have the cost function shown in Figure 17.7(a), then we can calculate the probabilities of the various costs occurring. Thus there is a cost of amount a if $u < \alpha - r/\sigma$ which has probability $\Phi(\alpha - r/\sigma)$. All the costs and probabilities are given in Table 17.4.

The total expected cost can thus be found by multiplying the costs by their probabilities and summing. This cost is a function of α, so α can be chosen to minimize the cost. Doing this leads to the equation

$$(a - b)\phi\left(\frac{\alpha - r}{\sigma}\right) + (b - c)\phi\left(\frac{\alpha}{\sigma}\right) + (c - d)\phi\left(\frac{\alpha + s}{\sigma}\right) = 0$$

where $\phi(u)$ is the ordinate of the standard normal curve. It will be seen that if the cost function is symmetrical with $a = d$, $b = c$ and $r = s$ the solution of the equation is $\alpha = 0$ as we would expect. For the non-symmetrical case with $b = c = 0$ the equation becomes

$$\phi\left(\frac{\alpha - r}{\sigma}\right) \Big/ \phi\left(\frac{\alpha + s}{\sigma}\right) = d/a$$

Table 17.4 Costs and probabilities corresponding
to Figure 17.7(a)

Cost	Range of u	Probability
a	$u < \dfrac{\alpha - r}{\sigma}$	$\Phi\left(\dfrac{\alpha - r}{\sigma}\right)$
b	$\dfrac{\alpha - r}{\sigma} \leqslant u \leqslant \dfrac{\alpha}{\sigma}$	$\Phi\left(\dfrac{\alpha}{\sigma}\right) - \Phi\left(\dfrac{\alpha - r}{\sigma}\right)$
c	$\dfrac{\alpha}{\sigma} \leqslant u \leqslant \dfrac{\alpha + s}{\sigma}$	$\Phi\left(\dfrac{\alpha + s}{\sigma}\right) - \Phi\left(\dfrac{\alpha}{\sigma}\right)$
d	$\dfrac{\alpha + s}{\sigma} < u$	$1 - \Phi\left(\dfrac{\alpha + s}{\sigma}\right)$

The implication of this is that the greater d/a the greater α as is to be expected.

The same procedure can be applied to cost functions such as those in Figure 17.7(b) and (c) and also where the costs depend not just on the magnitude of the error but on the values of x and $\hat{y}$. These methods tend to lead to rather involved equations, but they can provide a useful method of adjusting forecasts to meet more realistic cost criteria.

Example 2

Let us now examine an example where the criteria of interest is not a cost but a probability. Suppose we wish our stage II forecast to be such that the probability of the future observation being more than 10 per cent. out is as small as possible. That means to say that we want

$$P = \text{Prob } (x_t < 0.9\,\hat{y}_t \text{ or } x_t > 1.1\,\hat{y}_t)$$

to be small. In general we would require to get $\hat{y}_t$ to minimize an expression of the form

$$P = \text{Prob } (x_t < b\hat{y}_t \text{ or } x_t > a\hat{y}_t)$$

From the distribution of the stage I error we can find the distribution of x_t conditional on $\hat{x}_t$. For example, x_t might be distributed as $N(\hat{x}_t, \sigma^2)$. Thus we can find P as a function of both $\hat{y}_t$ and $\hat{x}_t$. We may then choose $\hat{y}$ to minimize P. If we denote the distribution of x_t given $\hat{x}_t$ by $f(x_t \mid \hat{x}_t)$ this exercise gives

$$\frac{f(b\hat{y}_t \mid \hat{x}_t)}{f(a\hat{y}_t \mid \hat{x}_t)} = \frac{a}{b}$$

as the relation giving minimum P. As an example, if the stage I forecast

Table 17.5 Examples of stage II forecasts

$a = 1.1, \; b = 0.9$

$\hat{x}_t$	σ^2	$\hat{y}_t$	$\hat{x}_t$	σ^2	$\hat{y}_t$
10	10	16.20	100	1	100.10
20	10	24.15	100	4	101.58
50	10	51.93	100	5	110.24
100	10	100.99	100	50	208.38
500	10	500.20	100	100	366.76

$a = 2.0, b = 0.5$

$\hat{x}_t$	σ^2	$\hat{y}_t$
100	1	80.00
100	50	80.46
100	200	81.80
100	1000	88.37
100	2000	95.48
100	4000	107.50

errors are normally distributed with zero mean and variance σ^2, left-hand side of the above equation becomes

$$\exp - \frac{1}{2\,\sigma^2} \{ (b\hat{y}_t - \hat{x}_t)^2 - (a\hat{y}_t - \hat{x}_t)^2 \}$$

Hence we find

$$\hat{y}_t = \left[\hat{x}_t + \left\{ \hat{x}_t^2 + 2\sigma^2 \log \frac{a}{b} \frac{a+b}{a-b} \right\}^{\frac{1}{2}} \right] \bigg/ (a+b)$$

Some illustrative values of $\hat{y}_t$ are given in Table 17.5. It will be noted that if $\sigma^2/\hat{x}^2$ is small, i.e. we have good forecasts in percentage mean square error terms, then the above result approximates to

$$\hat{y} = 2\hat{x}/(a+b)$$

Thus for $a + b = 2$, as in the symmetrical example we started with, $\hat{y} = \hat{x}$. If, however, we allow for a higher percentage of error above $\hat{y}$, then a increases and so $\hat{y}$ moves below $\hat{x}$.

References

Bates, J. M., and Granger, C. W. J. (1959). The combination of forecasts. *Op. Res. Quarterly*, **20**, 451—468.

Bunn, D. W. (1975). A Bayesian approach to the linear combination of forecasts. *Op. Res. Quarterly*, **26**, 325—330.

Granger, C. W. J. (1969). Prediction with a generalised cost of error function. *Op. Res. Quarterly*, **20**, 199—207.

Newbold, P., and Granger, C. W. J. (1974). Experience with forecasting univariate time series and the combination of forecasts. *J. Roy. statist. Soc. A*, **137**, 131—146.

Theil, H. (1958). *Economic forecasts and Policy*. North Holland Publishing Co., Amsterdam.

Theil, H. (1965). *Applied Economic forecasting*. North Holland Publishing Co., Amsterdam.

Zeilner, A. (1963). Decision rules for economic forecasting. *Econometrica*, **31**, 111—120.

Chapter 18

Problems in practice

18.1 Introduction

The aim of this chapter is to discuss some problems that arise when forecasting is used in practice. In this discussion we will not attempt anything like a full coverage of the field. Instead, we will examine a number of selected examples. The examples have been chosen to illustrate one main theme. This theme is that it is incorrect to regard forecasting as an exercise that can be done on its own without reference to the use to which we put the forecasts. All too frequently this is exactly what is done. Having produced a forecast it is regarded as a number that can be used for any situation where a forecast is required. We might say that in such a case the forecast and application are 'decoupled' from each other. Such decoupling often occurs where a firm has a 'forecasting program' on its computer used simply to generate forecasts which are then given to anyone who needs them. There are several possible sources of trouble in adopting this type of approach to forecasting. These are discussed and illustrated in this chapter. In summary, incorrect decoupling of forecast and application can lead to:

(a) the use of forecasts derived from inappropriate forecasting criteria;
(b) the use of inappropriate forecasting parameters;
(c) the occurrence of forecasts that may be very accurate as forecasts but are nonetheless of little value when used in the application;
(d) the occurrence of self-defeating forecasts;
(e) the creation by the forecast of unfortunate side effects in the situation;
(f) the occurrence of self-validating forecasts.

There are situations where decoupling is quite possible and satisfactory. There are others where a 'partial decoupling' is possible. In these latter situations an initial forecast is obtained without reference to the application and is later modified to suit a particular application. These topics were dealt with in Chapter 17 on 'two-stage forecasting'.

The alternative to decoupling and partial decoupling is the integration of the forecasting aspect with other aspects of a particular application to produce a 'system' for doing the job required.

In this chapter the aim is to illustrate some of the phenomena referred to in (a) to (f) above.

18.2 Practical forecasting criteria — a stock control example

The following example is based on a paper by Trigg and Pitts (1963). The reason for selecting this example is that it illustrates how forecasting can be closely linked to an application and also how a forecasting parameter can be chosen to optimize a criterion related directly to the application.

The basic problem of stock control is that of giving the best service, to those who wish to draw on stock, at the lowest investment in stock. To give the best service it is necessary to make the probability of running out of stock as small as seems reasonable. In this example a probability of 0.025 (2½ per cent) has been chosen. To achieve this the total stock in hand and expected deliveries to the stock should be greater than the forecast demand by a safety factor which is called the 'safety stock'. If the demand from the customer each week has been $\ldots, x_{t-1}, x_t$, then the forecast demand for next week using exponential smoothing is

$$\tilde{x}_t = (1-a) \sum_{r=0}^{\infty} a^r x_{t-r} \tag{1}$$

One way of choosing safety stock is to let it be several weeks demand plus some constant amount q. How many weeks are used will depend on the time taken between ordering for stock and receiving these orders. To simplify this example, we assume this time to be one week and hence the safety stock is one week's demand plus q, i.e. $\tilde{x}_t + q$. Suppose that the amount ordered from the producer at the end of week t is P_t and the amount of stock at that time is S_t. The change in stock over the last week is $S_t - S_{t-1}$, which must equal the deliveries based

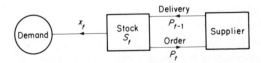

Figure 18.1   Model for the stock control example

on the previous week's order P_{t-1} minus the amount drawn x_t. Hence

$$S_t - S_{t-1} = P_{t-1} - x_t \tag{2}$$

It is also necessary that our latest order for replenishment P_t must be such that with our present stock S_t it must meet next week's demand, forecast as $\tilde{x}_t$, and provide for the safety stock. Hence

$$P_t + S_t = \tilde{x}_t + (\tilde{x}_t + q) \tag{3}$$

Changing the suffix to $t-1$ and subtracting gives

$$P_t - P_{t-1} = S_{t-1} - S_t + 2(\tilde{x}_t - \tilde{x}_{t-1}) \tag{4}$$

Substituting from equation (2) in (4) gives

$$P_t = x_t + 2(\tilde{x}_t - \tilde{x}_{t-1}) \tag{5}$$

Substituting this in equation (3) again gives

$$S_t = -x_t + 2\tilde{x}_{t-1} + q \tag{6}$$

Equation (6) thus relates the stock held to the latest demand, the previous forecast and the size of the safety stock.

We now consider two possible situations. In the first we assume that the demands are independent and follow a normal distribution with variance σ^2. As x_t is independent of x_{t-1}, the variance of S_t, σ_s^2, is given by

$$\sigma_s^2 = \sigma^2 + 4 \operatorname{Var}(\tilde{x}_{t-1})$$

using equation (6), which is

$$\sigma_s^2 = \sigma^2 \left(1 + 4\frac{1-a}{1+a}\right) \tag{7}$$

using equation (1). To achieve our 2½ per cent probability of 'stock out', we would make our safety stock equal to $2\,\sigma_s$. We can then make this small by making σ_s small; this would be done by making a large and, in fact, if we could assume a global constant mean model we would use the ordinary mean as the forecast in place of $\tilde{x}_t$. However, in the usual stock control situation such an assumption is often unrealistic. It is more reasonable to expect trends to occur in the data, at least locally. The lag of the forecast x_t behind an upward trend $x_t = \beta t$ would soon lead to the safety stock being used up. To avoid this, we add to the safety stock an amount D_m which is the maximum depression in stock produced by such a trend.

Table 18.1 shows some illustrative values of the various quantities in this situation calculated by using equations (2) and (3). It is seen from this table that the system is such that after the fourth week the orders delivered overtake the trend in sales, so that the depletion in stock

Table 18.1 The effect of trend on stock

Week t	0	1	2	3	4	5	6
Stock at $t-1, S_{t-1}$	10	10	9.00	8.40	8.12	8.10	8.28
Orders delivered P_{t-1}	0	0	1.40	2.72	3.98	5.18	6.34
Sales x_t	0	1	2.00	3.00	4.00	5.00	6.00
Stock at t, S_t	10	9	8.40	8.12	8.10	8.28	8.62
Orders made P_t	0	1.4	2.72	3.98	5.18	6.34	7.48
Forecast $\tilde{x}_t$	0	0.2	0.56	1.05	1.64	2.31	3.05
Depletion D_t	0	1.0	1.60	1.86	1.90	1.72	1.38

reaches a maximum at week 4. To find the depletion D_t in stock at week t we define D_t by

$$D_t = S_0 - S_t \tag{8}$$

which, by equation (6), taking the initial stock S_0 as just q, gives

$$D_t = x_t - 2\tilde{x}_{t-1}$$

Substituting the model for x_t in $\tilde{x}_t$ gives

$$\tilde{x}_t = \beta t - a\beta(1 - a^t)/(1 - a)$$

Thus on substituting and sorting out we find

$$D_t = \beta\{2 - (1 - a)t - 2a^t\}/(1 - a) \tag{9}$$

Figure 18.2 shows a rough sketch of this type of curve. It is seen that there is a maximum depletion D_m in stock after a time t_m (to the nearest integer), but that after this the stock in fact starts to increase and the system overcompensates for the trend. In theory this looks bad, but in practice the type of trend we are considering here is simply a local phenomenon. If it appears to have become a permanent part of the situation, the forecasting formula would be changed to a trend-correcting form. The intuitive explanation of the form of D_t is that initially in the transient state, the forecast does not follow the trend, so

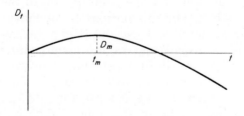

Figure 18.2   Depletion in stock — sketch of shape of curve

the orders do not keep up with the demand. After a time the forecast increases at the same rate as the trend but with a fixed lag behind it. Though in this steady state the forecasts are lower than the sales, the occurrence in the ordering rule of equation (5) of the term 2 $(\tilde{x}_t - \tilde{x}_{t-1})$, which has expectation 2β, means that the orders will on average be for more than next week's sales and thus, in the steady state, there will be a continual increase in stocks. The calculation of D_m from D_t is best done by evaluating D_1, D_2, etc., until the largest value is found. If there were no random variation in sales and we could reasonably guarantee that no local trends would have a slope greater than β, then a safety stock greater than D_m would ensure that no 'stock out' is likely to occur.

If we now allow for both random variation and the possibility of a trend, we see that it is reasonable to set the safety stock at $2\,\sigma_s + D_m$. Referring to equations (7) and (9) it is clear that both these quantities depend on the choice of a, if a is increased then σ_s decreases and D_m increases, and vice versa. Thus there is, in fact, a value of a for which the safety stock is minimum. This value, for example, is approximately 0.95 for $\sigma/\beta = 10$. We thus have here an example where a forecasting parameter is chosen, not with reference to getting best forecasts in any mean square error sense but with reference to the particular application in which the forecast is used.

18.3 The forecast as part of a system — a stockholding example

In this section we will look a little more at the idea of stockholding. We saw in section 18.2 that the choice of forecasting parameters can be related to the quantities required of a stock control system. However, if one did aim at optimum forecasts for the system one could treat the forecasting and stock control as separate exercises. Very often in business situations the very data one uses for forecasting is affected by somebody else's efforts at stock control. In section 18.2 we looked at a stock control application from the point of view of the person concerned with the operation of the stock control system. Let us now look at that same situation from the point of view of the person or firm who supplied the stock. For this person the quantity P_t is the order that comes to him at time t and goes out later as his sales. For the sake of concreteness le us regard him as the manufacturer and then the stocks S are those held by the retailer and x_t refers to the retailer's sales. If the public is only going to buy, say, 1,000 items from the retailer it is clearly best for the manufacturer to only manufacture 1,000 items. However, to be in this happy state he would have to produce a perfect forecast of the values. Unfortunately, he is usually not able to do this and in fact as he does not have access to the retailer's data he cannot forecast x at all. The only information he has is the data of his sales which is the sequence of the P values. One would like to think that this gave him relevant information about future sales to the public so that he can plan his manufacturing accordingly, Let us examine, therefore, the relation between the sales to the public and the manufacturer's sales. Assuming the same simple situation envisaged in section 18.2 we have equation (5) of that section as

$$P_t = x_t + 2(\tilde{x}_t - \tilde{x}_{t-1})$$

Let us notice a number of different facets of this equation.

(a) Suppose we are using exponential smoothing with

$$\tilde{x}_t = a\tilde{x}_{t-1} + (1-a)x_t$$

we then have

$$P_t = (3 - 2a)x_t - 2(1-a)\tilde{x}_{t-1}$$

If for simplicity we assume that x_t and $\tilde{x}_{t-1}$ are independent, then the variance of P_t is given by

$$\mathrm{Var}(P_t) = (3 - 2a)^2 \ \mathrm{Var}(x_t) + 4(1-a)^2 \ \mathrm{Var}(\tilde{x}_{t-1})$$

It is sufficient here to simply note that the coefficients of $\mathrm{Var}(x_t)$ is bigger than one. If $a = 0.8$ it is $(1.4)^2$. Thus the manufacturer's sales have a greater variability than the retailer's and it is consequently more difficult for him to produce forecasts of future requirements. The reason for this additional variation lies in the use of a safety stock whose size depends on forecast sales.

(b) Let us now consider the case where sales show a steady rate of growth or decline and that the retailer makes a simple forecast of the form:

$$\tilde{x}_t = cx_t$$

Thus if he expects a 5 per cent growth rate he would use $c = 1.05$. In this case we have

$$P_t = x_t + 2c(x_t - x_{t-1})$$

Thus if x is in fact increasing $P_t > x_t$, irrespective of c, and if x is decreasing $P_t < x_t$. The consequence of this is that if sales are increasing the manufacturer will overestimate, in his forecasts, the total amount to be sold ultimately to the public If sales are decreasing he will underestimate. The amount of overestimation or underestimation will increase with the rate of change of x. An example of this type of behaviour is given in the paper by Kuchn and Day (1963), where the term 'acceleration effect' is used to describe the phenomena.

(c) Suppose now we consider what happens if there is an oscillation in the sales x. For simplicity let us suppose that the sales x follow the sequence $b + m$, $b - m$, $b + m$, $b - m$, ... With the forecasting formula of (b) a little arithmetic will show that P would follow the sequence $b + (1 + 4\,c)m$, $b - (1 + 4\,c)m$, $b + (1 + 4\,c)m$, ... Thus the height, or amplitude, of the oscillation has increased by a factor of $(1 + 4\,c)$. Thus the man who always assumes that next month will be the same as this, $c = 1$, will have magnified the oscillation by a factor of five. There is a further feature here worth noticing, and that is that as P_t is actually sold at time $t + 1$ the manufacturer will be delivering $b - (1 + 4\,c)m$ to the retailer when the retailer is delivering $b + m$ to the ultimate customer. This type of occurrence can cause the retailer to worry and play safe by putting up the value of c next time he orders. Thus in our example there is a change in amplitude and also a lag between the peaks of x and those of P.

In (a) to (c) above we have considered a very special example with a lead time of one and some particularly simple safety stock and forecasting rules. These examples, however, illustrate some very widely occurring phenomena. Figure 18.3 summarizes the phenomena discussed. Often there is more than one middle-man in the system. For example, in areas of clothing manufacture the clothes are bought from a shop, which carries a stock. The shop buys from a wholesaler, who carries a stock. The wholesaler buys from the manufacturer, who carries

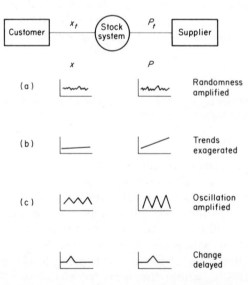

Figure 18.3   Some effects of stockholding

a stock of clothes and also a stock of material. This manufacturer may buy from a materials manufacturer who may in turn buy from a manufacturer of the appropriate fibres. Each, of course, will have his own stocks. Any change in the sales of clothes in the shop will slowly work its way back through the system, being magnified each time. In one classic example of this the sales to customers changed by something like 5 per cent. It took 6 months for this change to work its way through the chain and the change was amplified to 50 per cent by the time it reached the producer of raw materials.

A more involved study of the single-stock situation is given in the paper by Leicester (1963), which looks at the sales to and by retailers in the shoe trade. This has a number of complicating features. A retailer's subjective expectations of the state of the market influence his orders from the manufacturer. In this situation the delivery time is a varying quantity also depending on the state of the market. The amount of safety stock is dependent on this delivery time or rather on the retailer's estimate of what it will be. A consequence of the amplifications and variable lags produced by this system is that when the sales of shoes by retailers show a steady upward movement, manufacturers' sales show a strong cyclical oscillation of booms and then recessions.

Having been warned of the effects that stocks may have on orders from stockholders, what should the forecaster do with such order data? There are a number of actions open to him. He may seek to obtain data on sales to the ultimate customer which will provide the forecasts necessary for manufacturing. If this is not possible, he must take his forecast orders as information upon which the decision as to how much to manufacture will be based. To reach this decision he will need to develop a realistic model of the total system that generated his data. If the system is fairly simple, he may even be able to use his order data to estimate the sales-to-customer data, e.g. in (b) above knowledge of the sequence of the P values provides information about the sequence of the x values.

18.4 Forecasts and crystal balls — a scheduling example

A very common error in the use of forecasts is to treat the forecast in the same way as one would treat the true future observation, if only one had a crystal ball in which to see it. However, the number that comes out of the crystal ball is a known constant, whereas the number that one obtains from one's efforts in statistical forecasting is essentially the value taken by a random variable. This distinction between a constant and a random variable must be clearly realized if forecasts are to be correctly used in practice. The object of the following section is to emphasize this point by considering examples in which the replacing of a constant value by a forecast can completely change the situation.

Let us begin by considering an idealized example to show that a forecast treated as a 'crystal ball' forecast can lead to a different result to the one which would be obtained if it is correctly treated as a random variable. The problem is to decide whether to increase plant capacity by one or two units ($C = 1$ or $C = 2$), assuming that the future additional demand, D, will be for either one or two units.

Model

Possible demand D	1 unit	2 unit
Average weekly profit		
if $C = 1$	10	10
if $C = 2$	5*	20

(*penalty for surplus plant)

(a) Suppose the crystal ball forecast demand is $D = 1$ unit. If $C = 1$ then the profit is 10, if $C = 2$ the profit is 5. Hence the decision is $C = 1$ for maximum profit.

(b) Suppose that the forecast, $\hat{D} = 1$, is based on estimated probabilities:

$$\text{Prob}(D = 1) = 0.6$$

$$\text{Prob}(D = 2) = 0.4$$

The expected profit

if $C = 1$ is $10 \times 0.6 + 10 \times 0.4 = 10$

if $C = 2$ is $\ 5 \times 0.6 + 20 \times 0.4 = 11$

(multiplying the profits by the probability of showing these profits). Hence the decision is $C = 2$ for maximum profit. Thus knowledge about the random aspects of the situation leads to a different decision being obtained from the forecast $\hat{D} = 1$.

Let us now consider a less idealized example in more detail. A very common practical problem is the choice of the number of 'servers' required to give a good service to 'customers'. Common examples of this are the number of cash desks in a supermarket and the number of telephone operators manning a switchboard. We consider here the specific problem of scheduling how many operators are required to man a switchboard. To answer this type of question, we require first of all information about the criteria of good service and, secondly, a model of how the particular queueing system functions. As a measure of service we take the probability that a caller can get straight through to an operator. For a numerical example we will say that service is satisfactory if this probability is greater than 0.99 or, conversely, that the probability of the caller being delayed is less than 0.01. The factors effecting the service the caller obtains are:

(a) the number of operators N;
(b) the statistical properties of the length of time taken by an operator to deal with a call;
(c) the statistical properties of the way in which calls arrive;
(d) the way in which calls queue up when all operators are engaged;
(e) the total load on the switchboard, θ.

In this example θ is measured in units of 100 seconds of contact between operators and callers. Over many years of study the actual behaviour of telephone systems has been described very successfully by a number of fairly simple statistical models. The model used for the numerical calculations in this section is the Erlang C model which is one of the simplest of these models. The principles of the discussion, however, apply to all models. Morse (1957) explains these models in detail.

Table 18.2 Maximum loads for Prob (delay) $\leqslant 0.01$. Units are 100 seconds of operator contact per half-hour for the Erlang C model

Number of operators M	1	2	3	4	5	6	7	8
Maximum θ, (θ_M)	0.1	2.6	7.7	14.4	22.5	31.5	41.2	51.7
Number of operators M	9	10	15	20	25	30	40	50
Maximum θ, (θ_M)	62	73	133	197	265	335	479	627

If we use the Erlang C model the probabilities of callers being delayed can be calculated for any given number of operators and any specified load. Table 18.2 gives the type of result obtained from such calculations. It is seen, for example, that, with four operators and the criteria of good service given above, the highest load that could be dealt with is 14.4 units. When the tables are used in practice, they are in fact used the other way round. For example, the question might be 'How many operators are required to give good service when the load is 30 units?'. From the table the answer is seen to be six operators.

In practice the load on telephone switchboards varies considerably during the day and also between the days of the week. Figure 18.4 shows the type of variation that occurred in one case. It is thus necessary to have different numbers of operators available at different times during the day and during the week. If we knew, for example, what the mean load was going to be for each half hour during the week,

284

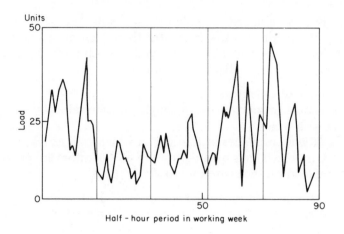

Figure 18.4   Variation in load on switchboard

we could use Table 18.2 to schedule the number of operators to man the switchboard throughout the week, so that callers are assured their 99 chances out of 100 of getting directly to an operator. Thus if we had our crystal ball to see what next week's load would be, there would be no real problem. However, what we in fact possess are forecasts based on observations on previous week's loads. What is usually done in this situation is to take the forecast, say 25 units for Monday 9.00—9.30am, and use it in the tables to decide how many operators are required, in this case six, exactly as though it were the true value. But our forecast will probably be wrong and the true value may turn out to be 35 units for which we should have had seven operators. Thus we have one less operator than is needed to give good service, and this is a direct result of our forecasting error.

An immediate question that arises is whether we can use the statistical properties of the forecasting error to obtain the probabilities of being in error in the numbers of operators. This can in fact be done either by a detailed study of the queueing model or by an approximate method. We will indicate briefly the approach of the approximate method as it, or some modification of it, can be used to study the effects of forecast errors in any situation where the forecasts are used in conjunction with tables such as Table 18.2.

The idea of the calculation is indicated by Figure 18.5. It is assumed that the forecast $\hat{\theta}$ is distributed about the true θ with a normal distribution of variance σ^2. It is also assumed that over small ranges of 3 or 4 times σ the table can be approximately represented by a set of equal intervals of width C on the axis of θ, each interval corresponding to a different number of operators. It is clear from Figure 18.5 that if we knew the position of θ we could find immediately the probabilities

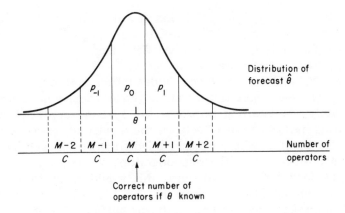

Figure 18.5   Effects of forecast errors on reading a table

Table 18.3 Approximate probabilities
of being K servers in error, Prob(K =
$-k$) = Prob($K = k$), for the Erlang C
model

K	$r = C/\sigma$		
	0.5	0.75	1.0
0	0.195	0.286	0.369
1	0.173	0.221	0.241
2	0.121	0.102	0.067
3	0.066	0.028	0.008
4	0.029	0.005	0.000
5	0.010	0.001	0.000
6	0.003	0.000	0.000
7	0.001	0.000	0.000
8	0.000	0.000	0.000

of being any specified number of servers in error. For example, the
probability of being understaffed by one would be p_{-1}, the area of the
appropriate region of the distribution curve. However, the position of θ
in the interval C is not known. It seems reasonable that it is equally
likely to be anywhere in the interval. If it is reasonable to make this
assumption we can obtain, using the idea of conditional and marginal
distributions, the probabilities of being, say, understaffed by one,
without assuming θ known. Table 18.3 gives the results of this type of
calculation. The probabilities of having the wrong numbers of servers is
seen to depend on the ratio $r = C/\sigma$. This is a measure of how far the
distribution of the forecast is spread over the intervals of the table.
Thus with $r = 1$ about 99 per cent of the distribution covers four lines

of the table. In the type of application in which this problem arose this corresponds to quite a good forecast, but yet it gives a probability of 0.24 of being understaffed by one. With larger forecast variability considerably worse errors are seen to occur.

As the probabilities of being understaffed and over-staffed are equally likely, one might think that no great damage has been done. Unfortunately, overstaffing improves the system very little whereas understaffing has disastrous effects on the quality of service given. For example, for the case where we scheduled six operators and had a load of 35 units, which required seven operators, the delay probability is 0.03. If the load had been 50 units, this would go up to 0.07. Thus though the probabilities of error in the numbers of operators is balanced between understaffing and overstaffing, the effect on the probability of delay is to increase it above the value of 0.01 used in the original table.

The fact that the load is forecast rather than known completely changes the meaning of the tables. Table 18.4 gives approximate values for the actual delay probabilities when forecast values of load are used in Table 18.2. The basis for the derivation of Table 18.4 was the use of the probabilities of Table 18.3 together with information from queueing theory on the effects of under- and overstaffing. It is clear from Table 18.4 that if one has a small load for which two of three operators are adequate, then large forecasting errors can lead to very poor service. One way to improve service is obviously to improve the forecasts. Whether or not this can be done, the obvious answer to the difficulties that are produced by the use of forecasts in the tables is to use the forecasts in a different way to the way one would use the true load if it were known. One simple method of doing this would be to use, in the tables, not the forecast but an adjusted forecast. The adjusted forecast would be larger than the original forecast, thus cutting down on the probability of understaffing and putting up the pro-bability of overstaffing; this in turn will decrease the actual probability of delays occurring.

We have discussed above a situation where a forecast, $\hat{\theta}$, is used to choose between a number of alternatives, $M = 1$, $M = 2$, etc. The basic objective is to use the forecast to make the best decision M in terms of some criteria, here the delay. We have discussed this in some detail as it is an example of a very common type of forecasting situation.

In summary, the lesson of this example is that we cannot treat forecasts in the same way as we would treat the true values. The effect of the random variation in forecasts can, however, be studied and appropriate adjustments made once its effects are understood.

Before leaving this example, it is perhaps worth noting the possibility of extending to other applications the use of the type of information shown in Figure 18.5. Suppose, for example, that a study of a forecast

Table 18.4 Approximate actual probabilities
of delay when using tables

(a) Probability of delay given in table = 0.01

Number of operators	$r = C/\sigma$		
	0.5	0.75	1.0
2	0.116	0.074	0.046
3	0.077	0.041	0.024
4	0.059	0.027	0.015
5	0.044	0.020	0.012
6	0.036	0.016	0.010
7	0.027	0.012	0.008
8	0.024	0.011	0.008
9	0.020	0.010	0.007
10	0.017	0.009	0.007

(b) Probability of delay given in table = 0.1

Number of operators	$r = C/\sigma$		
	0.5	0.75	1.0
2	0.185	0.149	0.123
3	0.172	0.132	0.106
4	0.151	0.113	0.092
5	0.138	0.102	0.083
6	0.131	0.096	0.079
7	0.123	0.091	0.076
8	0.113	0.084	0.071
9	0.108	0.081	0.070
10	0.103	0.078	0.068

errors leads to an estimate of the distribution of future demand for some product given the forecast, e.g. that shown in Figure 18.6. Suppose that on the basis of this certain production capacity, P, is invested in. If the demand lies in the region A there will be idle capacity under normal working, if in B overtime will be required, if in C additional production facilities will be required and if in D the demand will not be met. If the costs of these alternatives can be analysed then, the optimum production capacity P can be chosen. A paper by Thorneycroft, Greener and Patrick (1968) details such an analysis.

18.5 The forecast as part of a system — self-validating and self-defeating forecasts

There are a number of situations where forecasts turn out to be very accurate or completely wrong, not because of the skill or otherwise of

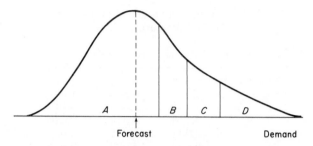

Figure 18.6   Consequences of deviation from forecast

the forecaster but because of the nature of the situation being dealt with. Let us briefly consider some examples of these effects.

(a) When a new product is launched, sales figures may show a nice steady increase. One is tempted then to forecast a continuing increase. There have been cases where this has been done only to find a very sharp levelling off at a later stage. The explanation has been that there has been a potential market for say 1,000 sales per week, but when the item was first introduced the production capacity was only geared to 500 a week. The production capacity has then been increased steadily. Thus until the 1,000 items/week figure was reached the sales equalled the production and forecasts based on the trend were naturally correct.

(b) The use of forecasts in setting salesmen's targets produces some interesting phenomena. Suppose the forecasts were made a target and no incentive was provided for passing the target, though one was provided for reaching the target. In this case one would expect the salesman's sales to equal the target and hence the forecast. Alternatively, if there was a fixed pay up to the target and good bonuses for exceeding the target, then there is a fair chance that the forecast will be exceeded. In essence one really needs to distinguish in this situation between a forecast of the natural market, the target given the salesmen and the forecast of the future sales, allowing for the influence of the target and the pay structure.

(c) In (b) we have a situation where the value forecast actually influences the future sales. Another example would be where the sales staff takes one look at the disasterous forecast of sales and takes drastic action to change the situation. In this case any resulting forecast error is not a measure of bad forecasting but partly of the success or otherwise of the actions taken.

(d) An interesting example is given by Forrester (1961), which, though there based on a simulation experiment, certainly happens in practice. This is the situation where an initial use of a seasonal forecasting method either creates a seasonal that need not exist or exaggerates one that does exist. Thus at the peak period forecast, advertising is at its maximum and the salesmen are working all hours. The result is that sales are naturally at a peak and will probably be higher than forecast. Next year's seasonal forecast thus has a tendency to show a greater seasonal effect than this year's. As a seasonal forecasting method will forecast seasonal elements even if none exist in the data, it is possible by the above policy to induce a seasonal sales where none need exist. Once again we have an undesirable feature produced by using the forecast in a situation without due regard to the decision nature or complexity of the situation.

(e) Another example is related to oscillation features in situations where production is geared directly to forecast sales. It can happen that the sort of magnification that was seen in section 18.3(b) and (c) allows the usual random variation in the forecast sales to cause larger swings in production. A consequence of this can be that at some periods production cannot meet demands. Thus sales are artificially lower than they should be or than they were forecast to be. We are thus misled in future forecasts by the type of phenomena referred to in (a).

Once again the message of these examples is that forecasting cannot be treated in isolation. It must be considered in relation to the practicalities of the situation for which the forecast is needed.

18.6 Practical forecasting — a postscript

In the previous sections we have looked at particular problems arising in the practice of forecasting. Let us finish this chapter, and this book, by taking a brief look at some of the main issues that face the forecaster in developing a practical forecasting system. As we are concerned with statistical forecasting, we can leave aside aspects of administration and implementation and concentrate on those questions that are relevant to the use of statistical models and methods for obtaining and using forecasts in practice. Each question will be discussed briefly and reference is made to the sections of this book that are most relevant.

(a) Why is the forecast required?

This is not a statistical question, but a knowledge of the answer is essential in providing a framework in which to consider the more statistical aspects of forecasting. Questions raised on the appropriate criteria to use (e.g. Chapter 3 and sections 17.5 and 18.2) or on the level of complexity of the methods appropriate (e.g. Chapters 12 and 14) can only be answered once the main objectives are clear.

(b) What information and resources are available or obtainable?

It is no good having a superb model for a situation if in practice the relevant data are unobtainable or unreliable (Chapter 13). Similarly, it is no good introducing a sophisticated forecasting programme on the computer if nobody in the firm understands its basic assumptions and principles. The forecasting models, methods and application need to be related to the real situation in which they are to be used.

(c) What model should be used?

We have emphasized the concepts of structure (section 1.3) and of models (Chapters 2 and 4 to 12) throughout this book. The problem of identifying the model to use is basic (Chapter 12) and it must be acknowledged that a lot depends on the experience and knowledge of the forecaster. The more complex the models in use the more difficult the task (e.g. Chapter 7 and Appendix A). An approach often adopted is to use a simple model to obtain forecasts and to seek to improve it by looking for structure in the forecast errors (Chapter 15). It is sometimes found that a reasonable forecast can be obtained from a relatively simple intuitively chosen model and that improvements on this are not financially worth the effort involved. Where it is worthwhile, one can increase the sophistication of the models or find other ways of improving the forecasts without changing the basic model (Chapter 17).

(d) How should I regard the model?

Throughout this book we have used the concepts of global and local models (e.g. section 2.3). These ideas emerge from the consideration of stability as underlying the approach of statistical forecasting. If we can assume that we have correctly identified the model, that it is stable or that, if it does undergo a sudden change, we can identify and use the new model rapidly, then a global use of our model is clearly the best approach. If, however, we judge the situation to be subject to slow evolution in time, either on the basis of past data or knowledge of the environment in which the data will arise, then we can regard the model as an approximation that is fitted locally for forecasting purposes (e.g. Chapters 4 and 5).

(e) What method should be used?

Having decided on the model and the approach to its use, the method of obtaining a forecast is usually fairly clear, as we have illustrated in Chapters 4 to 11 and 14. A general discussion of methods was given in Chapter 12. Often with experience we will seek to extend and modify our initial methods (section 14.2 and chapters 17 and 18).

(f) How are the results of forecasting to be assessed and controlled?

In this chapter we have emphasized the essential relatedness of the forecast and its application. Thus to assess the forecasts we need to look at their value in terms of the application as well as in the statistical terms (Chapters 3, 15, 17 and 18). Often, of course, these two are intimately related. When we turn from a general investigation of our methods and start to develop a practical forecasting system, then we need methods of control of the day-to-day running of the system (Chapter 16).

References

Forrester, J. N. (1961). *Industrial Dynamics.* MIT Press, Cambridge, Mass.

Kuehn, A. A. and Day, R. L. (1963). The acceleration effect in forecasting industrial shipments. *Journal of Marketing,* **27**, 25—28.

Leicester, C. S. (1963). The cause of the shoe trade cycle. *British Boot and Shoe Institute,* **11**, 307—316.

Morse, P. M. (1957). *Queues, Inventories and Maintainance,* 1967 edn. John Wiley and Sons, New York.

Thorneycroft, W. T., Greener, J. W. and Patrick, H. A. (1968). Investment decisions under uncertainty and variability. *Op. Res. Quarterly,* **19**, 143—160.

Trigg, D. W. and Pitts, E. (1963). The optimal choice of the smoothing constant in an inventory control system. *Op. Res. Quarterly,* **13**, 287—298.

Appendix A

Some terminology and definitions
for stochastic processes

The aim of this appendix is to define some of the terms that have been used in previous chapters, particularly Chapter 7. It also seeks to provide some additional ideas that will be of use when some of the more advanced references are consulted.

A discrete stochastic process is a process that generates a sequence of observations, $\{x_i\}$, in time. Figure A.1 represents a particular sequence from such a process, often called a 'realization' of the process. The probability density function of x_i at a specific time i is given by

$$f(x_i)dx_i = \text{Prob}\{x_i \leqslant x_i \leqslant x_i + dx_i\}$$

i.e. it is the probability that the sequence passes through the small window, dx_i, in an imaginary wall built at time i. Similarly, if we imagine two such windows at times i and j, we can obtain a joint probability density, $f(x_i, x_j)$, as being defined from the probability of the process passing through both windows, dx_i and dx_j, as in Figure A.1(b). If we know that at time i the sequence has the value x_i, then the probability of it passing through the window dx_j is $f(x_j \mid x_i)dx_j$; this is the conditional probability.

If $f(x_j \mid x_i)$ does not depend on the value x_i, we say that the two random variables x_i and x_j are independent (Figure A.1c). If all variables $\{x_i\}$ at all times are independent of each other, then the whole process is called an independent stochastic process or sometimes 'white noise'. The sequence $\{\epsilon_t\}$ used throughout this book is formally defined as such a process.

If all the distributions of one, two or more variables depend only on the relative positions of the 'walls', and thus are not effected by any change in time origin, then the process is said to be stationary. The global use of the models in Chapter 7 essentially assumed that the processes being dealt with were stationary. The local use of the models

294

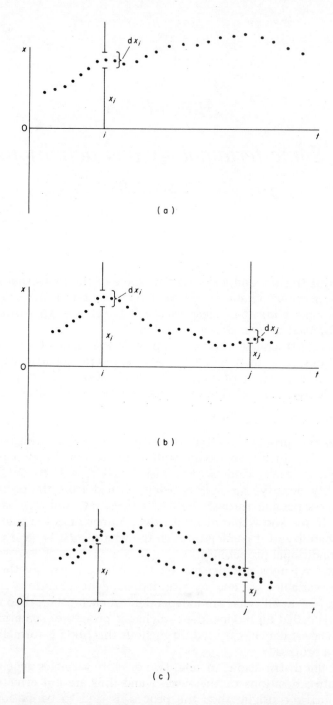

Figure A.1 Illustration for probability distribution
functions of stochastic processes

is based on the belief that often real processes are only locally stationary and are non-stationary over long periods of time.

We can define expectations for the above distributions as in Chapter 3, Table 3.5. Thus we can define:

$E(x_i)$ as giving the mean value of the process at time i

$E(x_j \mid x_i)$ as giving the mean value of the process at time j, given that it took the value of x_i at time i

$\text{Var}(x_i)$ giving the variance of the process at time i

We can also define some further quantities: e.g. the autocovariance of x_i and x_j,

$$\text{cov}(x_i, x_j) = E[\{x_i - E(x_i)\}\{x_j - E(x_j)\}]$$

This quantity measures the relationship between x_i and x_j. To see the way in which it does this we need to modify it slightly by defining the autocorrelation function of x_i and x_j by

$$\rho(x_i, x_j) = \text{Cov}(x_i, x_j) / \sqrt{(\text{Var}(x_i)\,\text{Var}(x_j))}$$

This quantity takes the value 1 if $x_i = x_j$, i.e. if x_i and x_j are perfectly linearly related, the value -1 if $x_i = -x_j$ and the value 0 if x_i and x_j are independent of each other. Thus $\rho(x_i, x_j)$ is a measure of linear relationship obtained by standardizing the autocovariance. Notice that

$$\text{Cov}(x_i, x_i) = \text{Var}(x_i)$$

If the process is stationary, $\text{Cov}(x_i, x_j)$ and $\rho(x_i, x_j)$ will depend only on the time difference $j - i$ and not on the actual times. Thus putting $j = i + k$ we can simplify our notation for a stationary process:

$$\text{Cov}\,(x_i, x_{i+k}) = C(k)$$

$$\rho\,(x_i, x_{i+k}) = \rho(k)$$

$$\text{Var}(x_i) = C(0)$$

so,

$$\rho(k) = C(k)/C(0)$$

The plot of $\rho(k)$ against k is called the correlogram.

By way of example, consider the first-order autoregressive process

$$x_t = \phi x_{t-1} + \epsilon_t \quad (|\phi| < 1)$$

if this is re-expressed as

$$x_t = \epsilon_t + \phi \epsilon_{t-1} + \phi^2 \epsilon_{t-2} + \ldots$$

296

it is straightforward to show that

$$E(x_t) = 0$$

$$\text{Var}(X_t) = (1 + \phi^2 + \phi^4 + \ldots)\sigma^2 = \sigma^2/(1 - \phi^2)$$

$$E(x_t x_{t+k}) = \phi^k(1 + \phi^2 + \phi^4 + \ldots)\sigma^2 = \sigma^2\phi^k/(1 - \phi^2)$$

Hence,

$$\text{Cov}(x_t, x_{t+k}) = \sigma^2\phi^k/(1 - \phi^2)$$

and

$$\rho(x_t, x_{t+k}) = \phi^k$$

None of these depend on t.

To develop the autocorrelation idea a little further, consider a third-order autoregressive process (see section 7.4):

$$x_t = \phi_1 x_{t-1} + \phi_2 x_{t-2} + \phi_3 x_{t-3} + \epsilon_t$$

If we multiply both sides by x_{t-1} and take expectations, we get

$$C(1) = \phi_1 C(0) + \phi_2 C(1) + \phi_3 C(2)$$

Dividing by $C(0)$ gives

$$\rho(1) = \phi_1 + \phi_2 \rho(1) + \phi_3 \rho(2)$$

If we multiply instead by x_{t-2} and repeat the process, we finish with

$$\rho(2) = \phi_1 \rho(1) + \phi_2 + \phi_3 \rho(1)$$

Finally, if we multiply by x_{t-3} instead of x_{t-2} and repeat, we obtain

$$\rho(3) = \phi_1 \rho(2) + \phi_2 \rho(1) + \phi_3$$

Thus we obtain three simultaneous equations relating the ϕ and the autocorrelations. These equations are called the Yule—Walker equations and are the model equations that correspond to the normal equation that we referred to in section 7.4. If we estimate $\rho(1)$, $\rho(2)$ and $\rho(3)$ from the data, we will obtain estimates of ϕ_1, ϕ_2 and ϕ_3. The estimate of ϕ_3, $\hat{\phi}_{33}$, is of particular importance, since if $\hat{\phi}_{33}$ is close to zero it suggests that the model should have order less than three. $\hat{\phi}_{33}$ is called the estimated third partial autocorrelation coefficient. Thus, in general, the quantity $\hat{\phi}_{kk}$ may be obtained by solving the Yule—Walker equations assuming a kth-order autoregressive model. The sequence $\hat{\phi}_{11}$, $\hat{\phi}_{22}, \ldots$ is used to decide on the order of the autoregressive process (see Table 12.2).

A process for which $E(x_t)$ and $\rho(x_t, x_{t+k})$ do not depend on t, but which may not be totally stationary in other properties, is said to be 'wide sense' stationary. If a process is stationary, it will automatically be wide sense stationary. The converse, though obviously not true in general, does hold if all the distributions in the process are normal, either univariate or multivariate. The normal distribution, sometimes called the Gaussian distribution, plays a major role in the theory of stochastic processes. This is because it has a number of valuable properties in theory and occurs fairly frequently in practice. In summary these are as follows.

(a) If the process is normal, all the distributions $f(x_i, x_j)$, $f(x_i)$, $f(x_j \mid x_i)$, etc., will be normal.
(b) Any linear sum of the process observations will be normal. For example, the exponentially weighted moving average obtained by exact exponential smoothing will be normally distributed.
(c) The three quantities $E(x_i) = \mu$, $\mathrm{Var}(x_i) = \sigma^2$ and $\rho(x_i, x_{i+k}) = \rho(k)$ completely define the stationary normal process.
(d) If $\rho(k) = 0$ $(k = 1, 2, \ldots)$, then the process is the independent normal process.
(e) The function of $x_t, x_{t-1}, \ldots$ that gives the forecast with smallest mean square error is in fact a linear combination of past observations.

Appendix B

Proof that conditional expectation gives minimum MSE forecasts

We here show that the forecast with the minimum mean square error is simply the expected value of the future x_{t+k}, given that we already know the values $x_1, x_2, \ldots, x_t$. This conditional expectation we denote by

$$E(x_{t+k} \mid x_1, \ldots, x_t)$$

Let us start by asking what constant $\tilde{x}_{t,k}$ we would use to forecast x_{t+k} to obtain the minimum mean square error. To find this we simply expand the mean square error and choose $\tilde{x}_{t,k}$ to give the minimum; thus

$$\text{MSE} = E(x_{t+k} - \tilde{x}_{t,k})^2$$

$$= E(x_{t+k}^2) - 2\tilde{x}_{t,k}E(x_{t+k}) + \tilde{x}_{t,k}^2$$

Differentiating with respect to $\tilde{x}_{t,k}$ and equating to zero gives

$$\tilde{x}_{t,k} = E(x_{t+k}) \tag{1}$$

and the mean square error for this $\tilde{x}_{t,k}$ is

$$\text{MSE} = E\{x_{t+k} - E(x_{t+k})\}^2 = \text{Var}(x_{t+k})$$

We thus obtain the best constant forecast by taking $\tilde{x}_{t,k}$ equal to the expected value of the future x_{t+k}.

To obtain a realistic forecast of x_{t+k} we must obviously use our data $x_1, \ldots, x_t$, which for now we regard as a sequence of random variables. Suppose we write the forecast as

$$\tilde{x}_{t,k} = \tilde{x}_{t,k}\,(x_1, \ldots, x_t)$$

We now want to find the *function* $\tilde{x}_{t,k}\,(x_1, \ldots, x_t)$ which gives the minimum mean square forecast error, so we want to minimize

$$\text{MSE} = E\{x_{t+k} - \tilde{x}_{t,k}(x_1, \ldots, x_t)\}^2$$

Now an ordinary expectation can be written in terms of conditional expectations. If we denote an expectation with respect to a random variable z by $E_z(\cdot)$, then a property of expectations is that the expectation with respect to two variables y and z can be written as

$$\underset{(y,z)}{E}(\cdot) = \underset{z}{E}\{\underset{y}{E}(\cdot \mid z)\}$$

Thus we first hold z constant and find the expectation with respect to y. This will give an answer which depends on z. We now find the expectation of this quantity with respect to z. Applying this idea to finding the mean square error we can rewrite the mean square error as

$$\text{MSE} = \underset{\mathbf{x}}{E}[\underset{x_{t+k}}{E}\{(x_{t+k} - \tilde{x}_{t,k}(x_1, \ldots, x_t))^2 \mid x_1, \ldots, x_t\}]$$

where the inner expectation is the conditional expectation given $x_1, \ldots, x_t$, denoted by $\mathbf{x}$, and the outer expectation is with respect to the joint distribution of $\mathbf{x}$. Now conditional on $x_1, \ldots, x_t$, the forecast $\tilde{x}_{t,k}(x_1, \ldots, x_t)$ is simply a constant, $\tilde{x}_{t,k}$, so the inner expectation could be written as

$$\underset{x_{t+k}}{E}[\{x_{t+k} - \tilde{x}_{t,k}\}^2 \mid \mathbf{x}]$$

We can choose $\tilde{x}_{t,k}$ to minimize this using the result obtained above, so the constant $\tilde{x}_{t,k}$ is the expectation of x_{t+k}, only now it is the expectation conditional on $\mathbf{x}$, i.e.

$$\tilde{x}_{t,k} = E\{x_{t+k} \mid \mathbf{x}\}$$

This choice must obviously minimize the total mean square error expression, as it is only the inner expectation that relates the past, $x_1, \ldots, x_t$, to the future, x_{t+k}. We thus arrive at the conclusion that the minimum mean square error forecast of a future observation, x_{t+k}, is the expectation of that observation given the data, $x_1, \ldots, x_t$, available at present, i.e.

$$\tilde{x}_{t,k}(x_1, \ldots, x_t) = E\{x_{t+k} \mid x_1, \ldots, x_t\}$$

Again using the results obtained for the constant forecast, it is clear that the minimum mean square error obtained using $\tilde{x}_{t,k}(x_1, \ldots, x_t)$ is

$$\text{MSE} = \underset{\mathbf{x}}{E}\{\text{Var}(x_{t+k} \mid x_1, \ldots, x_t)\}$$

That $\tilde{x}_{t,k}(x_1, \ldots, x_t)$ is unbiased follows from studying the error, which is

$$e_{t+k} = x_{t+k} - \tilde{x}_{t,k}(x_1, \ldots, x_t)$$

Taking conditional expectations of both sides gives

$$E(e_{t+k} \mid x) = E(x_{t+k} \mid x) - E\{\tilde{x}_{t,k}(x_1, \ldots, x_t) \mid x\}$$
$$= \tilde{x}_{t,k}(x_1, \ldots, x_t) - \tilde{x}_{t,k}(x_1, \ldots, x_t)$$

so

$$E(e_{t+k} \mid x) = 0$$

But

$$E(e_{t+k}) = \underset{x}{E}\{ E(e_{t+k} \mid x)\} = \underset{x}{E}(0)$$

and so

$$E(e_{t+k}) = 0$$

and

$\tilde{x}_{t,k}(x_1, \ldots, x_t)$ is thus an unbiased forecast of $X_{t,k}$.

Appendix C

Two-way exponential smoothing

In the discussion of local models and discounted least squares we have always used weights that put the greatest emphasis on present data. From the forecasting point of view this is the logical thing to do. However, if we wish to explore the local variation in the fitted model over past time this is not so useful. Suppose, for example, we wish to estimate the local values of the model parameters at some past time T. In this case one logical set of weights would be

Time	0	1	2	...	$T-1$	T	$T+1$	...	t
Weight	a^T	a^{T-1}	a^{T-2}	...	a	1	a	...	a^{t-T}

If we used these in discounted least squares to fit the model, discounting into the future from T as well as the past, we would minimize

$$\sum_{r=-T}^{t-T} a \mid r \mid (\text{residual}_{T+r})^2$$

In general this is a more complicated exercise than we have considered for estimating present parameter values. If we are estimating the local mean, with large body data, the method leads to a very simple modification of exponential smoothing. We simply generate weighted sums by sweeping both ways through the data. Using arrows to denote the direction of sweep, let

$$\overrightarrow{S}_i = a\overrightarrow{S}_{i-1} + x_i \qquad \overrightarrow{S}_0 = x_0$$
$$\overleftarrow{S}_i = a\overleftarrow{S}_{i+1} + x_i \qquad \overleftarrow{S}_t = x_t$$
$$\overrightarrow{W}_i = a\overrightarrow{W}_{i-1} + 1 \qquad \overrightarrow{W}_0 = 1$$
$$\overleftarrow{W}_i = a\overleftarrow{W}_{i+1} + 1 \qquad \overleftarrow{W}_t = 1$$

Then the local estimate of μ_T is

$$\hat{\mu}_T = (\overrightarrow{S}_T + \overleftarrow{S}_T - x_T)/(\overrightarrow{W}_T + \overleftarrow{W}_T - 1)$$

Appendix D

A technical note on discounted least squares

In Chapters 5 and 6 the basic concepts of discounted least squares were introduced and some results quoted. The aim of this appendix is to derive the normal equations for the linear form of model used in several chapters of this book, see Gilchrist (1967). Using the notation of Chapter 6, consider the particular model with, for the sake of illustration, three parameters. Thus the fitted model will look like

$$y_{t-r} = b_0 + b_1 x_{1,t-r} + b_2 x_{2,t-r} + \hat{e}_{t-r}$$

where the x values are known regressor variables, $\hat{e}_{t-r}$ are the residuals and $t-r$ is the time r time units before the current time t. The criterion for discounted least squares

$$S = \sum_{r=0}^{t-1} a^r \hat{e}_{t-r}^2$$

and we require b_0, b_1 and b_2 to be the values that minimize this with a as the fixed discounting factor ($0 < a \leqslant 1$). Substituting for $\hat{e}$ gives

$$S = \sum_{r=0}^{t-1} a^r (y_{t-r} - b_0 - b_1 x_{1,t-r} - b_2 x_{2,t-r})^2$$

To minimize this we differentiate with respect to b_0, b_1 and b_2 and equate each expression to zero. Thus

$$\frac{\partial S}{\partial b_0} = -2 \sum_{r=0}^{t-1} a^r (y_{t-r} - b_0 - b_1 x_{1,t-r} - b_2 x_{2,t-r}) = 0$$

$$\frac{\partial S}{\partial b_1} = -2 \sum_{r=0}^{t-1} a^r x_{1,t-r} (y_{t-r} - b_0 - b_1 x_{1,t-r} - b_2 x_{2,t-r}) = 0$$

$$\frac{\partial S}{\partial b_2} = -2 \sum_{r=0}^{t-1} a^r x_{2,t-r} (y_{t-r} - b_0 - b_1 x_{1,t-r} - b_2 x_{2,t-r}) = 0$$

Sorting these terms out and dropping the summation limits for clarity

302

gives

$$b_0 \Sigma a^r + b_1 \Sigma a^r x_{1,t-r} + b_2 \Sigma a^r x_{2,t-r} = \Sigma a^r y_{t-r}$$

$$b_0 \Sigma a^r x_{1,t-r} + b_1 \Sigma a^r x_{1,t-r}^2 + b_2 \Sigma a^r x_{1,t-r} x_{2,t-r}$$
$$= \Sigma a^r x_{1,t-r} y_{t-r}$$

$$b_0 \Sigma a^r x_{2,t-r} + b_1 \Sigma a^r x_{2,t-r} x_{1,t-r} + b_2 \Sigma a^r x_{2,t-r}^2$$
$$= \Sigma a^r x_{2,t-r} y_{t-r}$$

These are the normal equations discussed in section 6.2. Notice that the three equations are equivalent to

$$\Sigma a^r \hat{e}_{t-r} = 0$$

$$\Sigma a^r x_{1,t-r} \hat{e}_{t-r} = 0$$

$$\Sigma a^r x_{2,t-r} \hat{e}_{t-r} = 0$$

The normal equations used in section 5.2.3 are obtained by a simple change in notation:

(a) b_0 becomes the estimate, $\hat{\mu}_t$, of the current mean.
(b) b_1 becomes the current estimate of slope $\hat{\beta}_t$ by letting x_{t-r} be simply $-r$, the time into the past.
(c) x_2 and b_2 are set at zero as there are only two parameters and two equations.
(d) We use x as the dependent variable instead of the y here.

We thus obtain

$$\hat{\mu}_t \Sigma a^r - \hat{\beta}_t \Sigma a^r r = \Sigma a^r x_{t-r}$$

$$\hat{\mu}_t \Sigma a^r r - \hat{\beta}_t \Sigma a^r r^2 = \Sigma a^r r x_{t-r}$$

Reference

Gilchrist, W. G. (1967). Methods of estimation using discounting. *J. Roy. Statist. Soc.*, **29**, 355–369.

Index

306